CHARLES K. CHUI
Texas A&M University

Multivariate Splines

SOCIETY FOR INDUSTRIAL AND APPLIED MATHEMATICS
PHILADELPHIA, PENNSYLVANIA   1988

All rights reserved. No part of this book may be reproduced, stored, or transmitted in any manner without the written permission of the Publisher. For information, write the Society for Industrial and Applied Mathematics, 1400 Architects Building, 117 South 17th Street, Philadelphia, Pennsylvania 19103-5052.

Copyright 1988 by the Society for Industrial and Applied Mathematics.

Library of Congress Catalog Card Number 88-61569.

ISBN 0-89871-226-2.

Printed by Capital City Press, Montpelier, Vermont.

Contents

v   PREFACE

1   CHAPTER 1.   Univariate Splines

        1.1   *B*-Splines and Truncated Powers on Uniform Mesh
        1.2   Univariate Spline Spaces
        1.3   Some Basic Properties of *B*-Splines
        1.4   *B*-Spline Series
        1.5   Computation of *B*-Splines

15   CHAPTER 2.   Box Splines and Multivariate Truncated Powers

        2.1   Box Splines
        2.2   Basic Properties of Box Splines
        2.3   Multivariate Truncated Powers
        2.4   Box Spline Series

27   CHAPTER 3.   Bivariate Splines on Three- and Four-Directional Meshes

        3.1   Dimension
        3.2   Locally Supported Splines
        3.3   Minimally and Quasi-Minimally Supported Bivariate Splines
        3.4   Bases and Approximation Order

41   CHAPTER 4.   Bivariate Spline Spaces

        4.1   A Classical Approach
        4.2   Quasi-Crosscut Partitions
        4.3   Upper and Lower Bounds on Dimensions
        4.4   Approximation Order
        4.5   Other Subspaces

57   CHAPTER 5.   Bézier Representation and Smoothing Techniques

        5.1   Bézier Polynomials
        5.2   Smoothness Conditions for Adjacent Simplices
        5.3   Smoothness Conditions for Adjacent Parallelepipeds
        5.4   Smoothness Conditions for Mixed Partitions

71   CHAPTER 6.   Finite Elements and Vertex Splines
      6.1   Vertex Splines
      6.2   Generalized Vertex Splines
      6.3   Polynomial Interpolation Formulas and Examples
      6.4   Applications

93   CHAPTER 7.   Computational Algorithms
      7.1   Polynomial Surface Display
      7.2   Discrete Box Splines
      7.3   The Line Average Algorithm
      7.4   Bézier Nets of Locally Supported Splines

113   CHAPTER 8.   Quasi-Interpolation Schemes
      8.1   The Commutator Operator
      8.2   Polynomial-Generating Formulas
      8.3   Construction of Quasi-Interpolants
      8.4   Neumann Series Approach

129   CHAPTER 9.   Multivariate Interpolation
      9.1   Interpolation by Polynomials
      9.2   Lagrange Interpolation by Multivariate Splines
      9.3   Cardinal Interpolation with Nonsingular ϕ
      9.4   Cardinal Interpolation with Singular ϕ
      9.5   Scaled Cardinal Interpolation

157   CHAPER 10.   Shape-Preserving Approximation and Other Applications
      10.1   Shape-Preserving Approximation by Box Spline Series
      10.2   Shape-Preserving Quasi-Interpolation and Interpolation
      10.3   Application to CAGD
      10.4   Reconstruction of Gradient Fields
      10.5   Applications to Signal Processing

171   APPENDIX.   A Computational Scheme for Interpolation

177   BIBLIOGRAPHY

Preface

The ten chapters in this monograph are a compilation of the material I presented at the Regional Conference on **Theory and Applications of Multivariate Splines** held at Howard University in Washington, D.C. during the week of August 10–14, 1987.

Although spline functions of one variable have important applications in many areas of scientific and engineering research, they are frequently insufficient to meet the need of the oncoming recent technological developments. For instance, to represent a mathematical model described by several parameters and to interpret higher dimensional data, functions of two or more variables are often required. Among these, the family of piecewise polynomial functions in more than one variable, which are usually called multivariate splines, is the most useful in many applications. Surprisingly, until the late seventies and early eighties, very little information on the theory and computational methods of these functions was known. Since then, the subject of multivariate splines has become a rapidly growing field of mathematical research. It is remarkable that hundreds of papers in this area have been written within a period of less than ten years. The objective of the present monograph is to introduce this fascinating subject which has a great potential in applications to a larger audience of scientists and engineers. With this in mind, our approach follows an elementary point of view that parallels the theory and development of univariate spline analysis. To compensate for the missing proofs and details, an extensive bibliography on the various aspects considered in this monograph has been included.

The support by the National Science Foundation through a grant to the Conference Board of Mathematical Sciences is gratefully acknowledged. It is a pleasure to thank Professor Daniel Williams for doing a superb job in organizing the conference and the participants for being so enthusiastic throughout all the lectures and discussion sessions. In particular, the stimulating discussions with I. Borosh, H. G. Burchard, H. Diamond, E. T. Y. Lee, W. R. Madych, R. Mohapatra, C. Prather, L. Raphael, T. C. Sun, and J. D. Ward are very much appreciated. I would also like to acknowledge the U. S. Army Research Office and the National Science Foundation for supporting my research in this area.

In preparation of the manuscript, I have been greatly benefited by discussions with Rick Beatson and Harvey Diamond, and have also received helpful comments from Carl de Boor, Martin Buhmann, Wolfgang Dahmen, Eugene Lee, Tom Lyche, Charles Micchelli, Amos Ron, and Larry Schumaker. To them I am very appreciative. In addition, I would also like to thank Harvey Diamond for providing the appendix of this monograph. His contribution helps in enhancing the numerical aspect of interpolation of gridded data. The final version of the manuscript was prepared while I was a visiting Erskine Fellow at the University of Canterbury, Christchurch, New Zealand. I am grateful to the members of the Mathematics department for their hospitality.

Finally, I am greatly indebted to Adrienne Morgan for typing most of the manuscript; to Rick Beatson, Guanrong Chen, and Ming-Jun Lai for spending long hours in helping with TeX; to Margaret Chui and Tian-Xiao He for assisting with the proof-reading; and to Laura Helfrich at SIAM for doing an excellent job in editing.

College Station   CHARLES K. CHUI

CHAPTER 1

Univariate Splines

A good understanding of the univariate (polynomial) spline theory and some familiarity with the available techniques are obviously important for learning this rapidly developing subject of multivariate (polynomial) splines. This chapter is intended to cover some of the basic topics in the classical theory of spline functions in one variable that have parallel (or orthogonal?) extensions to the multivariable setting. Since we will restrict our attention only to the essential topics that are somewhat related to the material that will be discussed in the next nine chapters, much of the beautiful theory and many important applications of univariate spline functions are not going to be covered. In particular, the approach that spline functions are considered as solutions of certain extremal problems will not be discussed here, since the multivariate analogue leads to another area of study, called "thin plate splines" (see the introduction to Chapter 9 and Duchon [101]). We refer the reader to the texts by de Boor [20], Schoenberg [180], and Schumaker [186] for further study of the classical theory and methods in univariate polynomial spline functions.

1.1. B-splines and truncated powers on uniform mesh. Let χ_A denote, as usual, the characteristic function of a given set A. Then the nth-order B-spline $B_n(x)$ on the uniform mesh is defined, inductively, as follows:

$$B_1(x) = \frac{1}{2}\left(\chi_{[-\frac{1}{2},\frac{1}{2}]}(x) + \chi_{(-\frac{1}{2},\frac{1}{2})}(x)\right)$$

and for $n = 2, 3, \cdots$,

$$B_n(x) = B_{n-1} * B_1(x)$$
$$= \int_{-\frac{1}{2}}^{\frac{1}{2}} B_{n-1}(x-t)dt.$$

The following properties of $B_n(x)$ can be verified easily by using mathematical induction. Throughout, $\pi_{n-1} = \pi_{n-1}^1$ will denote the space of all polynomials in one variable of order n, or degree at most $n-1$.

THEOREM 1.1.

(i) $B_n \in C^{n-2}(\mathbf{R})$.

(ii) *For any even integer* n, $B_n\big|_{[j,j+1]} \in \pi_{n-1}$, $j \in \mathbf{Z}$.

For any odd integer n, $B_n\big|_{[j-\frac{1}{2},j+\frac{1}{2}]} \in \pi_{n-1}$, $j \in \mathbf{Z}$.

(iii) $\operatorname{supp} B_n = [-\frac{n}{2}, \frac{n}{2}]$.

(iv) $B_n(x) > 0$ *for* $-\frac{n}{2} < x < \frac{n}{2}$.

(v) $\sum_j B_n(x-j) = 1$ *for all* x.

(vi) $\int_{-\infty}^{\infty} B_n(x)dx = 1$.

(vii) $\int_{-\infty}^{\infty} B_n(x)f(x)dx = \int_{[-\frac{1}{2},\frac{1}{2}]^n} f(t_1 + \cdots + t_n)dt_1 \cdots dt_n$
for all $f(x)$ *in* $C(\mathbf{R})$.

(viii) $B_n'(x) = B_{n-1}(x+\frac{1}{2}) - B_{n-1}(x-\frac{1}{2})$.

Substituting $f(x) = e^{-ixy}$ in (vii), we have the following corollary.

COROLLARY 1.1. *The Fourier transform of* $B_n(x)$ *is*

$$(1.1) \qquad \widehat{B}_n(y) = \left(\frac{\sin(y/2)}{y/2}\right)^n.$$

In the following we will use the notation

$$\Delta f(x) = f\left(x + \frac{1}{2}\right) - f\left(x - \frac{1}{2}\right),$$

and for $n = 1, 2, \cdots$,

$$\Delta^n f(x) = \Delta^{n-1}(\Delta f(x)).$$

Hence, another trivial consequence of (vii) in Theorem 1.1 is the following formula.

THEOREM 1.2.

$$(1.2) \qquad \int_{-\infty}^{\infty} B_n(x) g^{(n)}(x) dx = \Delta^n g(0).$$

We now recall the truncated power functions:

$$x_+ = \max(x, 0)$$

and

$$x_+^n = (x_+)^n, \quad n = 2, 3, \cdots.$$

By applying the Taylor formula with integral remainder and the identity:
$$(x-t)^{n-1} = (x-t)_+^{n-1} - (-1)^n(t-x)_+^{n-1},$$
we have the formula

(1.3) $$\Delta^n g(0) = \frac{1}{(n-1)!} \int_{-\infty}^{\infty} g^{(n)}(x) \Delta^n x_+^{n-1} dx.$$

Hence, by combining (1.2) and (1.3), we have the formula that expresses the B-spline $B_n(x)$ in terms of the truncated powers, namely:

THEOREM 1.3.

(1.4) $$B_n(x) = \frac{1}{(n-1)!} \Delta^n x_+^{n-1}.$$

1.2. Univariate spline spaces. Let $a = t_0 < \cdots < t_{m+1} = b$ and consider the space
$$\mathcal{S}_{\mathbf{t},n} = \{f \in C^{n-2}[a,b]: \; f|_{[t_i, t_{i+1}]} \in \pi_{n-1}, \; i = 0, \cdots, m\},$$
which is called the spline space of order n and with knot sequence $\mathbf{t} = \{t_i\}, i = 1, \cdots, m$. To study this space, let $f \in \mathcal{S}_{\mathbf{t},n}$ and
$$f\Big|_{[t_i, t_{i+1}]} = p_{i+1} \in \pi_{n-1}.$$
Then since
$$(p_{i+1} - p_i)^{(k)}(t_i) = 0$$
for $k = 0, \cdots, n-2$, we have

(1.5) $$p_{i+1}(x) = p_i(x) + c_i(x - t_i)^{n-1}$$

where c_i is a constant. That is, if $x \in [t_j, t_{j+1}]$, then

(1.6) $$f(x) = p_1(x) + \sum_{i=1}^{j} c_i(x - t_i)^{n-1}$$
$$= p_1(x) + \sum_{i=1}^{m} c_i(x - t_i)_+^{n-1},$$

where the second formulation of $f(x)$ is independent of the location of x. In other words, $S_{t,n}$ is a subspace of the linear span

$$\langle 1, \cdots, x^{n-1}, (x-t_1)_+^{n-1}, \cdots, (x-t_m)_+^{n-1} \rangle.$$

It is clear, however, that the collection

$$\mathcal{B} = \{1, \cdots, x^{n-1}, (x-t_1)_+^{n-1}, \cdots, (x-t_m)_+^{n-1}\}$$

of functions belongs to $S_{t,n}$ and is linearly independent on [a,b]. Hence, $\mathcal{B}$ is a basis of $S_{t,n}$. To give another basis consisting of only truncated powers, let us extend the knot sequence **t** to the left, with the extension also denoted by **t** for simplicity, namely:

$$\mathbf{t}: \quad t_{-n+1} < \cdots < t_{-1} < a = t_0 < \cdots < t_{m+1} = b.$$

Then it is clear that the collection

$$\mathcal{B}_t = \{(x-t_{-n+1})_+^{n-1}, \cdots, (x-t_m)_+^{n-1}\}$$

is also a basis of $S_{t,n}$. To produce another basis consisting of elements with smaller supports, we further extend the knot sequence **t** to the right, and again with the extension also denoted by **t**, yielding:

$$\mathbf{t}: \quad t_{-n+1} < \cdots < t_{m+1} = b < t_{m+1} < \cdots < t_{m+n}.$$

Now, we need the notation of the nth divided difference

$$[t_i, \cdots, t_{i+n}]f$$

to define the so-called normalized B-splines:

(1.7) $$N_{t,n,i}(x) = (t_{i+n} - t_i)[t_i, \cdots, t_{i+n}]_t (t-x)_+^{n-1}$$

where the divided difference is taken at the variable t. Since $N_{t,n,i}(x)$ is a linear combination of the functions

$$(t_i - x)_+^{n-1}, \cdots, (t_{i+n} - x)_+^{n-1},$$

it is clearly in $S_{t,n}$. The linear independence of the collection

$$\mathcal{B}_b = \{N_{t,n,-n+1}(x), \cdots, N_{t,n,m}(x)\}$$

of functions on [a,b] can be easily verified, and this implies that $\mathcal{B}_b$ is also a basis of $\mathcal{S}_{\mathbf{t},n}$. In other words, any spline function is a linear combination of B-splines. Hence, it is sometimes more convenient to consider a bi-infinite knot sequence:

$$(1.8) \qquad \mathbf{t}: \quad \cdots < t_{-k} < \cdots < t_k < \cdots$$

where $t_k \to \infty$ and $t_{-k} \to -\infty$ as $k \to \infty$, and consider the so-called *B-spline series*

$$(1.9) \qquad \sum_i c_i N_{\mathbf{t},n,i}(x) = \sum_{i=-\infty}^{\infty} c_i N_{\mathbf{t},n,i}(x)$$

where for each x, the infinite series is only a finite sum of at most n terms.

1.3. Some basic properties of B-splines. In the special case when $t_i = i \in \mathbf{Z}$, it turns out that

$$(1.10) \qquad N_{\mathbf{t},n,i}(x) = B_n\left(x - \frac{n}{2} - i\right)$$

where $B_n(x)$ is the B-spline discussed in §1.1. We list some of the most important properties of $N_{\mathbf{t},n,i}(x)$ in the following theorem.

THEOREM 1.4.
(i) supp $N_{\mathbf{t},n,i} = [t_i, t_{i+n}]$.

(ii) $N_{\mathbf{t},n,i}(x) > 0$ for $t_i < x < t_{i+n}$.

(iii) $\sum_i N_{\mathbf{t},n,i}(x) = 1$ *for all* x.

(iv) $\int_{-\infty}^{\infty} N_i(x) dx = (t_{i+n} - t_i)/n$.

(v)
$$\left[\frac{1}{t_{i+n} - t_i} N_{\mathbf{t},n,i}(x)\right] = \frac{x - t_i}{t_{i+n} - t_i}\left[\frac{1}{t_{i+n-1} - t_i} N_{\mathbf{t},n-1,i}(x)\right]$$
$$+ \frac{t_{i+n} - x}{t_{i+n} - t_i}\left[\frac{1}{t_{i+n} - t_{i+1}} N_{\mathbf{t},n-1,i+1}(x)\right]$$

for all x.

(vi) $N_{\mathbf{t},n,i}(x)$ *has minimal support.*

Note that (v) says that an nth-order B-spline can be obtained as a linear combination of two $(n-1)$st-order B-splines with the same knot sequence and that the relationship in (v) describes a convex combination. Hence, it not only gives a recurrence computational formula for the B-splines, but also indicates an interesting geometric property. We remark, however, that this computational scheme is carried out for each fixed x. In §1.5, we will introduce a method that gives an explicit formula for each polynomial piece of $N_{t,n,i}(x)$.

That the support of $N_{t,n,i}(x)$ is minimal means that if $s \in S_{t,n}$ is supported on a proper subinterval of supp $N_{t,n,i} = [t_i, t_{i+n}]$, then s must be identically zero. On the other hand, it is certainly clear that $N_{t,n,i}(x)$ is unique up to a multiplicative constant among all $s \in S_{t,n}$ with support equal to supp $N_{t,n,i}$.

1.4. B-spline series. The B-spline series (1.9) is very important since it gives a local representation of any spline function in the space $S_{t,n}$. Of course, if all the coefficients c_i are chosen to be 1, then the series sums to 1 everywhere as stated in (iii) of Theorem 1.4. This property of "partition of unity" is a very useful ingredient in the study of approximation by spline functions. It may be considered as a formula to locally generate constants. To generate other polynomials in π_{n-1} locally, we have the following identity due to Marsden [153].

THEOREM 1.5. *Let* $p \leq q$. *Then for* $t_p \leq x \leq t_{q+1}$,

$$(1.11) \qquad (y-x)^{n-1} = \sum_{i=p-n+1}^{q} \left[\prod_{\ell=1}^{n-1} (y - t_{i+\ell}) \right] N_{t,n,i}(x).$$

Consequently,

$$(1.12) \qquad x^{j-1} = $$

$$\sum_{i=p-n+1}^{q} (-1)^{j-1} \frac{(j-1)!}{(n-1)!} \left\{ D^{n-j} \left[\prod_{\ell=1}^{n-1} (y - t_{i+\ell}) \right] \Bigg|_{y=0} \right\} N_{t,n,i}(x) $$

for $j = 1, \cdots, n$.

From the fact that π_{n-1} can be locally generated, it is expected that the order of approximation of sufficiently smooth functions from the spline space $S_{t,n}$ is $O(h^n)$, where

$$h = \max_{i}(t_{i+1} - t_i).$$

In fact, not only is this true, but explicit formulas of spline approximants of f that give this optimal order can also be written in the form

$$\text{(1.13)} \qquad \sum_i \lambda_i(f) N_{\mathbf{t},n,i}(x)$$

where each λ_i is a linear functional with small local support. Such formulas are called quasi-interpolation formulas, first studied in de Boor [18] by using point evaluation linear functionals and later in de Boor and Fix [27], where the linear functionals λ_i involve function values as well as derivatives. In Lyche and Schumaker [151], derivatives are replaced by divided differences.

If we choose

$$\text{(1.14)} \qquad \lambda_i(f) = f\left(\frac{t_{i+1} + \cdots + t_{i+n-1}}{n-1}\right);$$

that is, $\lambda_i(f)$ is chosen to be the value of $f(x)$ at the point which is the average of the knots interior to the support of $N_{\mathbf{t},n,i}(x)$, then the formula (1.13) gives the so-called Schoenberg's "variation diminishing" spline approximant $(Vf)(x)$ of $f(x)$. Although this simple approximant only yields an approximation order of $O(h^2)$, it preserves certain geometric shape characteristics of the data function $f(x)$. This result can be verified by using yet another important property of spline functions, i.e., that the number of sign changes of Vf does not exceed that of f (cf. [20], [186]). Indeed, since it follows from (1.12) that

$$(Vg)(x) := \sum_i g\left(\frac{t_{i+1} + \cdots + t_{i+n-1}}{n-1}\right) N_{\mathbf{t},n,i}(x) = g(x)$$

for all $g \in \pi_1$, we may conclude that the number of sign changes of $Vf - g = V(f - g)$ does not exceed that of $f - g$ for all $g \in \pi_1$. By choosing $g = 0$, we have

$$f \geq 0 \implies Vf \geq 0 \ ;$$

by choosing g to be any constant, we have

$$f \uparrow \implies Vf \uparrow \ ;$$

and by choosing g to be any straight line, we have

$$f \text{ convex} \implies Vf \text{ convex}.$$

The spline series in (1.9) is also important in obtaining interpolants from $\mathcal{S}_{\mathbf{t},n}$. An important problem is, of course, to determine the admissible location of the set of sample points $\{x_i\}$,

$$\cdots < x_1 < x_2 < \cdots.$$

The following result due to Schoenberg and Whitney [182] completely solves this problem. Let $a \leq x_1 < \cdots < x_{m+n} \leq b$ and the knot sequence $\mathbf{t}$ be defined as in § 1.2. Then we have the following theorem.

THEOREM 1.6. *The matrix*

$$[N_{\mathbf{t},n,i}(x_j)], \quad 1 \leq i, j \leq m+n,$$

is nonsingular if and only if $t_i < x_i < t_{n+i}$ for all $i, i = -n+1, \cdots, m$.

In other words, for each i, the sample point x_i must lie in the interior of the support of $N_{\mathbf{t},n,i}$ in order to guarantee that the interpolation problem will be poised.

1.5. Computation of B-splines. Many algorithms are available for computing $N_{\mathbf{t},n,i}(x)$. We may classify them into three types. The first type allows the user to compute $N_{\mathbf{t},n,i}(x)$ for each fixed value of x from some recurrence relationship such as (v) in Theorem 1.4. Of course, if we wish to determine $N_{\mathbf{t},n,i}(y)$ for y different from x, the same procedure can be carried out once more. The second type is to give an efficient approximation scheme which is based on some different recurrence relationship. This type is useful for displaying curves very efficiently and the algorithms are usually developed by applying "subdivisions." Several such algorithms are available in the literature. In the following, we will discuss the so-called line average algorithm (cf. Cohen, Lyche, and Riesenfeld [79], and Dahmen and Micchelli [94], [96]), which is valid only for the uniform mesh.

Set

$$N_n(x) = B_n(x - \frac{n}{2})$$

which has support $[0, n]$ and knots at the integers $\mathbf{Z}$. Considering $N_n(x)$ as a spline of order n and with knots at $\frac{1}{p}\mathbf{Z}$, where p is any positive integer, we may write $N_n(x)$ as a linear combination of the B-splines $N_n(px - j)$, namely:

$$(1.15) \qquad N_n(x) = \sum_{j=-\infty}^{\infty} a_p^n\left(\frac{j}{p}\right) N_n(px - j)$$

for some constants $a_p^n\left(\frac{j}{p}\right)$. Taking the Fourier transform of both sides of (1.15), we have

(1.16) $$\left[\frac{1-e^{-ix}}{ix}\right]^n = \sum_{j=-\infty}^{\infty} a_p^n\left(\frac{j}{p}\right) \frac{1}{p} e^{-ij\frac{1}{p}x} \left[\frac{1-e^{-i\frac{1}{p}x}}{i\left(\frac{x}{p}\right)}\right]^n$$

where we have used (1.1) and the definition

$$N_n(x) = B_n\left(x - \frac{n}{2}\right).$$

If we set $z = e^{-i\frac{1}{p}x}$, then (1.16) becomes

(1.17) $$\frac{1}{p^{n-1}}\left(\frac{1-z^p}{1-z}\right)^n = \sum_{j=-\infty}^{\infty} a_p^n\left(\frac{j}{p}\right) z^j.$$

Multiplying the left by $(1-z^p)/(1-z)$ and the right by the equivalent expression $1 + \cdots + z^{p-1}$, we have:

$$\frac{1}{p^{n-1}}\left(\frac{1-z^p}{1-z}\right)^{n+1} = \sum_{j=-\infty}^{\infty} a_p^n\left(\frac{j}{p}\right) \sum_{\ell=0}^{p-1} z^{j+\ell}$$

$$= \sum_{k=-\infty}^{\infty} \left(\sum_{\ell=0}^{p-1} a_p^n\left(\frac{k-\ell}{p}\right)\right) z^k.$$

Equating this identity with (1.17) where n is replaced by $n+1$ yields the *line average algorithm*:

(1.18) $$a_p^{n+1}\left(\frac{j}{p}\right) = \frac{1}{p}\sum_{\ell=0}^{p-1} a_p^n\left(\frac{j-\ell}{p}\right)$$

which may be used to compute

$$a_p^n\left(\frac{j}{p}\right)$$

recursively for $n = 1, 2, \cdots$, with initial value

(1.19) $$a_p^1\left(\frac{j}{p}\right) = \begin{cases} 1 & \text{for } j = 0, \cdots, p-1 \\ 0 & \text{otherwise}. \end{cases}$$

10   CHAPTER 1

Now let us discuss a method for displaying the graph of any spline series

(1.20) $$S_n(x) = \sum_{j=-\infty}^{\infty} c_j N_n(x-j)$$

with order n and knots at $\mathbf{Z}$. It is easy to see that

(1.21) $$S_n(x) = \sum_{j=-\infty}^{\infty} d_p^m\left(\frac{j}{p}\right) N_n(px-j)$$

where

(1.22) $$d_p^m\left(\frac{j}{p}\right) = \sum_{k=-\infty}^{\infty} a_p^n\left(\frac{j}{p}-k\right) c_k.$$

Hence, by using the relationships (1.18), (1.19), and (1.22), we have the following so-called *line average algorithm* for computing the coefficients

$$d_p^n\left(\frac{j}{p}\right)$$

of the spline series (1.21).

THEOREM 1.7. *For any $m \in \mathbf{Z}$ and $0 \leq k \leq p-1$,*

(1.23) $$d_p^1\left(m + \frac{k}{p}\right) = c_m$$

for $k = 0, \cdots, p-1$, and

(1.24) $$d_p^n\left(m + \frac{k}{p}\right) = \frac{1}{p} \sum_{j=0}^{p-1} d_p^{n-1}\left(m + \frac{k-j}{p}\right),$$

where $n = 2, 3, \cdots$.

For $x = m + \frac{k}{p}$, where $m \in \mathbf{Z}$ and $0 \leq k < p$, we have, from (1.21), for $n > 1$,

$$S_n\left(m + \frac{k}{p}\right) = \sum_{j=1}^{n-1} d_p^m\left(m + \frac{k}{p} - \frac{j}{p}\right) N_n(j)$$

so that for large values of p, since $1 \leq j \leq n-1$, it follows, at least intuitively, that

$$S_n\left(m + \frac{k}{p}\right) \approx \sum_{j=1}^{n-1} d_p^n\left(m + \frac{k}{p}\right) N_n(j) = d_p^n\left(m + \frac{k}{p}\right).$$

In other words, $\{d_p^n\left(m + \frac{k}{p}\right)\}$ provides an approximation of $S_n(x)$ at the "dense" set $\{x = m + \frac{k}{p}: \ m \in \mathbf{Z}, 0 \leq k < p\}$.

Actually, with a shift of $n/2p$, the approximation has order $O(p^{-2})$, but cannot be $o(p^{-2})$ unless $S_n(x)$ is a linear spline (cf. Dahmen, Dyn, and Levin [86] and Cohen and Schumaker [81]).

The third type of algorithm is to give an explicit formulation for each polynomial piece of the B-spline, $N_{\mathbf{t},n,i}(x)$. In the following, we will introduce an algorithm given in Chui and Lai [67].

Let

$$\phi_k^n(x) = \binom{n}{k} x^k (1-x)^{n-k}$$

and

$$\phi_{j,k}^n(x) = \phi_k^n\left(\frac{x - t_j}{t_{j+1} - t_j}\right).$$

We will denote the restriction of $N_{\mathbf{t},n,i}(x)$ to $[t_j, t_{j+1}]$ by

$$P_{i,j}^n(x) = \sum_{k=0}^{n-1} a_k^{n-1}(i,j) \phi_{j,k}^{n-1}(x).$$

The set of coefficients $\{a_k^{n-1}(i,j)\}$ will be called the *Bernstein net* of the B-spline $N_{\mathbf{t},n,i}(x)$. We have the following result.

THEOREM 1.8. *Let* $a_k^{m-1}(i, i-1) = a_k^{m-1}(i, i+m) = 0$. *For each* $j = i, \cdots, i + m$ *and* $k = 0, \cdots, m-1$,

(1.25) $\qquad a_{k+1}^m(i,j) = a_k^m(i,j) + \dfrac{t_{i+j+1} - t_{i+j}}{t_{i+m} - t_i} a_k^{m-1}(i,j)$

$\qquad\qquad\qquad - \dfrac{t_{i+j+1} - t_{i+j}}{t_{i+m+1} - t_{i+1}} a_k^{m-1}(i+1, j-1)$

with initial conditions $a_0^m(i,i) = 0$ *and* $a_0^m(i,j) = a_m^m(i, j-1)$, $j = i+1, \cdots, i+m$.

In the special case of uniform mesh, say $t_k = k$, the formula (1.25) for computing the B-spline $N_n(x) = B_n(x - \frac{n}{2})$ is particularly simple, since by setting $i = 0$, it becomes:

(1.26) $\qquad a_{k+1}^m(0,j) = a_k^m(0,j) + \dfrac{1}{m}\left(a_k^{m-1}(0,j) - a_k^{m-1}(1, j-1)\right)$

with $a_0^m(0,0) = 0$, and $j, k = 0, \cdots, m-1$.

Hence, to compute $N_{m+1}(x)$ from $N_m(x)$, we first write down the Bernstein net for $\frac{1}{m}(N_m(x) - N_m(x-1))$, namely:

$$b_{jk}^{m-1} = \frac{1}{m}(a_k^{m-1}(0,j) - a_k^{m-1}(1, j-1)).$$

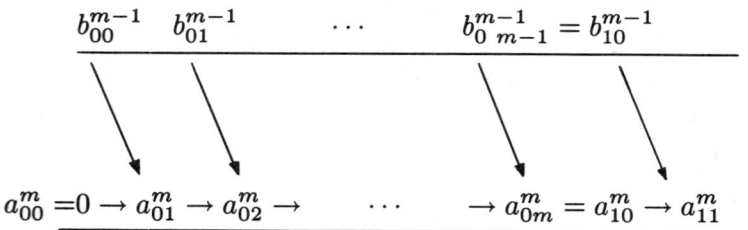

FIG. 1.1

Then the Bernstein net $a_{jk}^m = a_k^m(0,j)$ of $N_{m+1}(x)$ can be obtained by simple addition:

$$a_{j,k+1}^m = a_{jk}^m + b_{jk}^{m-1}, \quad k = 0, \cdots, m-1,$$

with $a_{j0}^m = a_{j-1,m}^m$ and the initial condition $a_{00}^m = 0$, where the index $j, j = 0, \cdots, m-1$, indicates the $(j+1)$st polynomial piece of $N_{m+1}(x)$. See Figure 1.1 above and Figure 1.2 below in computing the quadratic and cubic splinesfile. The computation of B-splines on a nonuniform mesh is more complicated. By using (1.25), we may find the Bernstein net of the quadratic and cubic splines as shown in Figure 1.3, where we have used the following notation:

$$\begin{cases} h_i = t_{i+1} - t_i \\ k_i = h_{i+1} + h_i = t_{i+2} - t_i \\ \ell_i = h_{i+2} + h_{i+1} + h_i = t_{i+3} - t_i \\ H_i = 1 - \dfrac{h_{i+2}^2}{k_{i+1}\ell_i} - \dfrac{h_{i+1}^2}{k_{i+1}\ell_{i+1}} \end{cases}$$

UNIVARIATE SPLINES

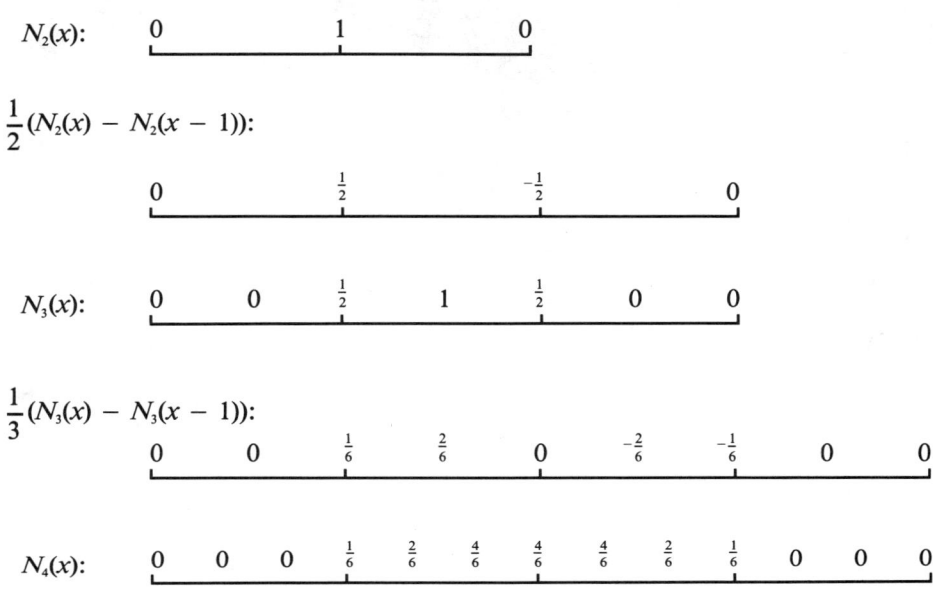

FIG. 1.2

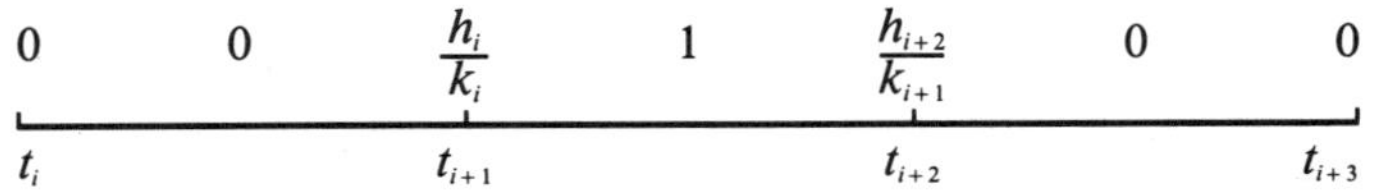

Quadratic spline $N_{\mathbf{t},3,i}$

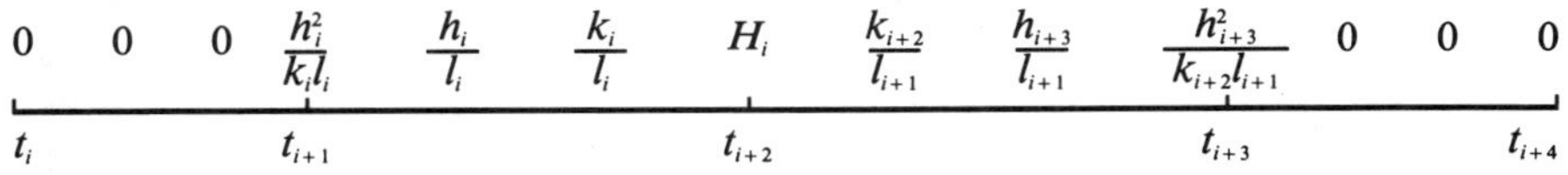

Cubic spline $N_{\mathbf{t},4,i}$

FIG. 1.3

CHAPTER 2

Box Splines and Multivariate Truncated Powers

The definition of B-splines and that of truncated powers with equally spaced knots in one variable will now be generalized to the multivariate setting. In this chapter, we will introduce the notion of box splines and multivariate truncated powers following the convolution procedure as has been done in defining univariate B-splines in §1.1. Some of the basic properties, including a recurrence relationship, relationships between box splines and truncated powers, approximation order, etc., will be discussed in this chapter.

2.1. Box splines. A natural generalization of the univariate B-spline on the uniform mesh to the multivariate setting is the so-called *box spline* introduced by de Boor and DeVore [25].
Let
$$X_n = \{\mathbf{x}^1, \cdots, \mathbf{x}^n\} \subset \mathbf{Z}^s \setminus \{\mathbf{0}\}$$
be a "direction set" with
$$\text{Span } X_n = \langle X_n \rangle = \mathbf{R}^s$$
and consider the *affine cube*:
$$[X_n] = [\mathbf{x}^1, \cdots, \mathbf{x}^n]$$
$$= \left\{ \sum_{i=1}^n t_i \mathbf{x}^i : -\frac{1}{2} \le t_i < \frac{1}{2}, i = 1, \cdots, n \right\}.$$

Since $\langle X_n \rangle = \mathbf{R}^s$, the s-dimensional volume of $[X_n]$, denoted by $\text{vol}_s[X_n]$, is positive.
Rearrange $\{\mathbf{x}^1, \cdots, \mathbf{x}^n\}$, if necessary, so that
$$\text{vol}_s[\mathbf{x}^1, \cdots, \mathbf{x}^s] > 0,$$
and we have the following definition of the box spline $M(\cdot|X_n)$ with direction set X_n.

DEFINITION. Set

$$M(\mathbf{x}|\mathbf{x}^1,\cdots,\mathbf{x}^s) = \begin{cases} \dfrac{1}{\text{vol}_s[\mathbf{x}^1,\cdots,\mathbf{x}^s]} & \text{if } \mathbf{x} \in [\mathbf{x}^1,\cdots,\mathbf{x}^s] \\ 0 & \text{otherwise}. \end{cases}$$

Then for $m = s+1, \cdots, n$, define, inductively,

$$(2.1) \qquad M(\mathbf{x}|\mathbf{x}^1,\cdots,\mathbf{x}^m) = \int_{-\frac{1}{2}}^{\frac{1}{2}} M(\mathbf{x}-t\mathbf{x}^m|\mathbf{x}^1,\cdots,\mathbf{x}^{m-1})dt$$

and set $M(\cdot|X_n) = M(\cdot|\mathbf{x}^1,\cdots,\mathbf{x}^n)$.

Example 2.1. For $s = 1$, if we set $\mathbf{x}^1 = \cdots = \mathbf{x}^n = 1$, then it is clear that $M(x|X_n) = B_n(x)$, the nth-order B-spline discussed in §1.1.

The following is a generalization of (vii) in Theorem 1.1.

THEOREM 2.1.

$$(2.2) \qquad \int_{\mathbf{R}^s} M(\mathbf{x}|X_n)f(\mathbf{x})d\mathbf{x} = \int_{[-\frac{1}{2},\frac{1}{2}]^n} f\left(\sum_{i=1}^n t_i\mathbf{x}^i\right) dt_1 \cdots dt_n$$

for all $f \in C(\mathbf{R}^s)$.

For $n = s$, by using the transformation

$$\mathbf{t} = (t_1, \cdots, t_s) \mapsto \mathbf{x} = \sum_{i=1}^s t_i \mathbf{x}^i, \quad \mathbf{t} \in [-\frac{1}{2},\frac{1}{2}]^s,$$

so that the Jacobian of the transformation has absolute value

$$|\det J| = \left| \det \begin{bmatrix} x_1^1 & \cdots & x_s^1 \\ \vdots & & \vdots \\ x_1^s & \cdots & x_s^s \end{bmatrix} \right| = \text{vol}_s[\mathbf{x}^1,\cdots,\mathbf{x}^s],$$

we have

$$\int_{[-\frac{1}{2},\frac{1}{2}]^s} f\left(\sum_{i=1}^s t_i\mathbf{x}^i\right) dt_1 \cdots dt_s = \int_{[\mathbf{x}^1,\cdots,\mathbf{x}^s]} f(\mathbf{x})|\det J|^{-1}d\mathbf{x}$$

$$= \int_{\mathbf{R}^s} M(\mathbf{x}|\mathbf{x}^1,\cdots,\mathbf{x}^s)f(\mathbf{x})d\mathbf{x}.$$

This allows us to use mathematical induction. Indeed, by using the definition of $M(\cdot|X_n)$ and the induction hypothesis consecutively, we obtain:

$$\int_{\mathbf{R}^s} M(\mathbf{x}|X_n) f(\mathbf{x}) d\mathbf{x}$$

$$= \int_{\mathbf{R}^s} \int_{-\frac{1}{2}}^{\frac{1}{2}} M(\mathbf{x} - t\mathbf{x}^n | \mathbf{x}^1, \cdots, \mathbf{x}^{n-1}) f(\mathbf{x}) dt d\mathbf{x}$$

$$= \int_{-\frac{1}{2}}^{\frac{1}{2}} \int_{\mathbf{R}^s} M(\mathbf{x}|\mathbf{x}^1, \cdots, \mathbf{x}^{n-1}) f(\mathbf{x} + t\mathbf{x}^n) d\mathbf{x} dt$$

$$= \int_{-\frac{1}{2}}^{\frac{1}{2}} \int_{[-\frac{1}{2},\frac{1}{2}]^{n-1}} f\left(\sum_{i=1}^{n-1} t_i \mathbf{x}^i + t\mathbf{x}^n\right) dt_1 \cdots dt_{n-1} dt$$

$$= \int_{[-\frac{1}{2},\frac{1}{2}]^n} f\left(\sum_{i=1}^{n} t_i \mathbf{x}^i\right) dt_1 \cdots dt_n.$$

This completes the proof of the theorem.

By using

$$f(\mathbf{x}) = e^{-i\mathbf{x}\cdot\mathbf{y}}$$

in the above theorem, we have the following corollary.

COROLLARY 2.1. *The Fourier transform of* $M(\cdot|X_n)$ *is*

$$(2.3) \qquad \widehat{M}(\mathbf{y}|X_n) = \widehat{M}(\cdot|X_n)(\mathbf{y}) = \prod_{i=1}^{n} \frac{\sin(\mathbf{y}\cdot\mathbf{x}^i/2)}{\mathbf{y}\cdot\mathbf{x}^i/2}.$$

This is a generalization of Corollary 1.1.

Remark. It is clear from (2.2) or (2.3) that the definition of $M(\cdot|X_n)$ is independent of the order of the vectors $\mathbf{x}^1, \cdots, \mathbf{x}^n$ in the direction set X_n. In fact, (2.2) or (2.3) may be used to define $M(\cdot|X_n)$ instead.

For this reason, we may consider the following important examples. Details will be considered in the next chapter. Let

$$\mathbf{e}^1 = (1,0) \quad \text{and} \quad \mathbf{e}^2 = (0,1).$$

Example 2.2. Let $\mathbf{x} = (x,y) \in \mathbf{R}^2$ and set

$$M_{tuvw}(x,y) = M(\mathbf{x}|X_{t+u+v+w})$$

18   CHAPTER 2

with

$$X_{t+u+v+w} = \{\underbrace{\mathbf{e}^1, \cdots, \mathbf{e}^1}_{t}, \underbrace{\mathbf{e}^2, \cdots, \mathbf{e}^2}_{u}, \underbrace{\mathbf{e}^1+\mathbf{e}^2, \cdots, \mathbf{e}^1+\mathbf{e}^2}_{v}, \underbrace{\mathbf{e}^2-\mathbf{e}^1, \cdots, \mathbf{e}^2-\mathbf{e}^1}_{w}\}.$$

In addition, set

$$M_{tu} = M_{tu00} \quad \text{and} \quad M_{tuv} = M_{tuv0}.$$

It is clear that $M_{tu}(x, y) = B_t(x)B_u(y)$ where $B_m(x)$ is the mth-order univariate B-spline. That is, M_{tu} is a tensor-product B-spline. In Figure 2.1, we give the *supports* and "grid lines" of these box splines.

The grid lines separate the "polynomial pieces" of the box splines where, of course, some of the grid lines may not be active. This conclusion is a consequence of the following theorem. Let us first remark that the tensor-product B-spline M_{tu} is a piecewise polynomial whose polynomial pieces are separated by a *rectangular partition*, the box spline M_{tuv} is a piecewise polynomial whose polynomial pieces are separated by a *three-directional mesh* (or *type-1 triangulation* which is obtained by drawing in all diagonals with positive slope), and M_{tuvw} is also a piecewise polynomial whose polynomial pieces are separated by a *four-directional mesh* (or *type-2 triangulation* which is obtained by drawing in both diagonals of each rectangle).

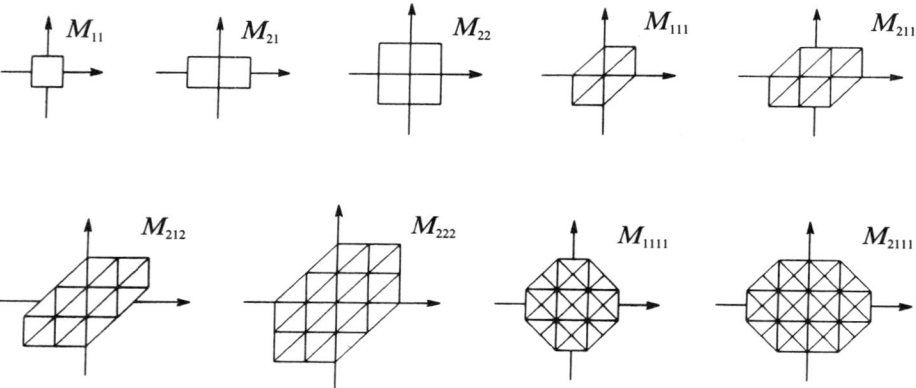

FIG. 2.1

THEOREM 2.2. *Let $X_n \subset \mathbf{Z}^s \backslash \{\mathbf{0}\}$ with $\langle X_n \rangle = \mathbf{R}^s$. Then the box spline $M(\cdot|X_n)$ has the following properties:*

(i) $\operatorname{supp} M(\cdot|X_n) = [X_n]$.

(ii) $M(\mathbf{x}|X_n) > 0$ *for* $\mathbf{x}$ *in the interior of* $[X_n]$.

(iii) *Set*

$$B_{X_n} = \left\{ \sum_{j=1}^{s-1} c_j \mathbf{x}^{i_j} + \sum_j b_j \mathbf{x}^{i'_j} : -\frac{1}{2} \leq c_j \leq \frac{1}{2}, b_j = \pm\frac{1}{2}, \right.$$
$$\left. 1 \leq i_1 < \cdots < i_{s-1} \leq n \right\}$$

where $\{i'_j\}$ denotes the complementary set of $\{i_j\}_1^{s-1}$ with respect to $\{1, \cdots, n\}$. (Clearly, $\operatorname{vol}_s B_{X_n} = 0$ and $B_{X_n} \subset [X_n]$.) Then the restriction of $M(\cdot|X_n)$ to each component of the complement of B_{X_n} is a polynomial of total degree $n - s$. (B_{X_n} is called the grid partition of the box spline $M(\cdot|X_n)$.)

(iv) *Let*

$$r(X_n) = \min\{\#Y : Y \subset X_n, \langle X_n \backslash Y \rangle \neq \mathbf{R}^s\} - 2.$$

Then $M(\cdot|X_n) \in C^{r(X_n)}(\mathbf{R}^s)$.

Note that (i) and (ii) follow from the definition, but (iii) and (iv) are not as trivial and the proof is omitted here (cf. de Boor and Höllig [29] and Höllig [133]).

NOTATION: S_d^r denotes the class of all piecewise polynomial functions in $C^r(\mathbf{R}^s)$ of total degree d.

Example 2.3. $M_{11} \in S_0^{-1}, M_{22} \in S_2^0, M_{111} \in S_1^0, M_{212} \in S_3^1, M_{222} \in S_4^2, M_{1111} \in S_2^1, M_{2111} \in S_3^1$.

2.2. Basic properties of box splines. Let $\mathbf{y} \in Y \subset \mathbf{R}^s \backslash \{\mathbf{0}\}, \mathbf{y} = (y_1, \cdots, y_s)$, where $\#Y < \infty$. We need the following notation:

(i) $D_\mathbf{y} = \sum_{i=1}^s y_i \frac{\partial}{\partial x_i}$

(ii) $D_Y = \prod_{\mathbf{y} \in Y} D_\mathbf{y}$

(iii) $\Delta_\mathbf{y} f = f(\cdot + \frac{\mathbf{y}}{2}) - f(\cdot - \frac{\mathbf{y}}{2})$

(iv) $\Delta_Y = \prod_{\mathbf{y} \in Y} \Delta_\mathbf{y}$.

The following result, which is a generalization of the univariate formula given in Theorem 1.2, can be obtained by using Theorem 2.1. Throughout, we will again assume that $X_n = \{\mathbf{x}^1, \cdots, \mathbf{x}^n\} \subset \mathbf{Z}^s \setminus \{\mathbf{0}\}$ with $\langle X_n \rangle = \mathbf{R}^s$.

THEOREM 2.3. *For any $f \in C^n(\mathbf{R}^s)$,*

$$(2.4) \qquad \int_{\mathbf{R}^s} M(\mathbf{x}|X_n) D_{X_n} f(\mathbf{x}) d\mathbf{x} = \Delta_{X_n} f(\mathbf{0}).$$

The univariate result in (viii) of Theorem 1.1 has the following generalization.

THEOREM 2.4. *For any j with $\langle X_n \setminus \{\mathbf{x}^j\} \rangle = \mathbf{R}^s$,*

$$(2.5) \qquad D_{\mathbf{x}^j} M(\cdot|X_n) = \Delta_{\mathbf{x}^j} M(\cdot|X_n \setminus \{\mathbf{x}^j\}).$$

The following recurrence formula was obtained by de Boor and Höllig [29].

THEOREM 2.5. *For any $\mathbf{x} \in \mathbf{R}^s$, write*

$$\mathbf{x} = \sum_{i=1}^n t_i \mathbf{x}^i$$

where each $t_i = t_i(\mathbf{x})$ is linear in $\mathbf{x}$. Then

$$(2.6) \qquad \begin{aligned} (n-s) & M(\mathbf{x}|X_n) \\ = \sum_{i=1}^n & \Big\{ (\tfrac{1}{2} + t_i) M(\mathbf{x} + \tfrac{1}{2}\mathbf{x}^i | X_n \setminus \{\mathbf{x}^i\}) \\ & + (\tfrac{1}{2} - t_i) M(\mathbf{x} - \tfrac{1}{2}\mathbf{x}^i | X_n \setminus \{\mathbf{x}^i\}) \Big\} \end{aligned}$$

whenever each $M(\cdot|X_n \setminus \{\mathbf{x}^i\})$ is continuous at $\mathbf{x} \pm \tfrac{1}{2}\mathbf{x}^i$, $i = 1, \cdots, n$.

The following result on partition of unity of the translates of a box spline can be easily derived from the definition.

THEOREM 2.6.

$$(2.7) \qquad \sum_{\mathbf{j} \in \mathbf{Z}^s} M(\cdot - \mathbf{j}|X_n) \equiv 1.$$

When it is necessary to change the direction set by a linear transformation, the following formula will be useful.

THEOREM 2.7. *Let A be a nonsingular s-dimensional square matrix such that $AX_n = \{A\mathbf{x}^1, \cdots, A\mathbf{x}^n\} \subset \mathbf{Z}^s$. Then*

$$(2.8) \qquad M(\mathbf{x}|X_n) = |\det A|M(A\mathbf{x}|AX_n).$$

2.3. Multivariate truncated powers. To generalize the truncated power $\frac{1}{(n-1)!}x_+^{n-1}$ to the multivariate setting, we will discuss the notion of *multivariate truncated powers* introduced by Dahmen [82].

Again, let $X_n = \{\mathbf{x}^1, \cdots, \mathbf{x}^n\} \subset \mathbf{Z}^s \backslash \{\mathbf{0}\}$ and consider the *affine cone*:

$$C(X_n) = C(\mathbf{x}^1, \cdots, \mathbf{x}^n)$$
$$= \left\{ \sum_{i=1}^n t_i \mathbf{x}^i : \ 0 \le t_i < \infty, i = 1, \cdots, n \right\}.$$

To define multivariate truncated powers, it is necessary to restrict the cone $C(X_n)$ to lie in some halfspace. More precisely, replacing $\mathbf{x}^i$ by $-\mathbf{x}^i$, if necessary, we will assume the first nonzero component of each $\mathbf{x}^i$ to be positive. An important consequence is that for each $\mathbf{x}$ in $\mathbf{R}^s$ and $i = 1, \cdots, n$, the interval

$$\{t > 0: \ \mathbf{x} - t\mathbf{x}^i \in C(X_n)\}$$

has finite length, so that the following formulation of multivariate truncated powers guarantees a finite value for each $\mathbf{x}$ in $\mathbf{R}^s$. On the other hand, this assumption does not change the definition of the box spline with direction set X_n, since

$$M(\cdot|\mathbf{x}^1, \cdots, \mathbf{x}^n) = M(\cdot|\pm\mathbf{x}^1, \cdots, \pm\mathbf{x}^n)$$

for arbitrary choices of the plus and minus signs. As before, rearrange $\{\mathbf{x}^1, \cdots, \mathbf{x}^n\}$, if necessary, so that

$$\text{vol}_s[\mathbf{x}^1, \cdots, \mathbf{x}^s] > 0.$$

DEFINITION. Set

$$T(\mathbf{x}|\mathbf{x}^1, \cdots, \mathbf{x}^s) = \begin{cases} \dfrac{1}{\text{vol}_s[\mathbf{x}^1, \cdots, \mathbf{x}^s]} & \text{if } \mathbf{x} \in C(\mathbf{x}^1, \cdots, \mathbf{x}^s) \\ 0 & \text{otherwise} \end{cases}$$

and for $m = s+1, \cdots, n$, define, inductively,

$$T(\mathbf{x}|\mathbf{x}^1, \cdots, \mathbf{x}^m) = \int_0^\infty T(\mathbf{x} - t\mathbf{x}^m|\mathbf{x}^1, \cdots, \mathbf{x}^{m-1})dt.$$

The multivariate truncated power with direction set X_n is defined by $T(\mathbf{x}|X_n) = T(\mathbf{x}|\mathbf{x}^1, \cdots, \mathbf{x}^n)$. It is obvious that supp $T(\cdot|X_n) = C(X_n)$.

Example 2.4. For $s = 1$, if we set $\mathbf{x}^1 = \cdots = \mathbf{x}^n = 1$, then it is clear that

$$T(x|X_n) = \frac{1}{(n-1)!} x_+^{n-1}.$$

Following the same proof as that of Theorem 2.1, we have the following theorem.

THEOREM 2.8. *For any continuous function $f(\mathbf{x})$ on $\mathbf{R}^s$ with compact support,*

(2.9) $$\int_{\mathbf{R}^s} T(\mathbf{x}|X_n) f(\mathbf{x}) d\mathbf{x} = \int_{[0,\infty)^n} f\left(\sum_{i=1}^n t_i \mathbf{x}^i\right) dt_1 \cdots dt_n.$$

Of course, it is clear from (2.9) that $T(\mathbf{x}|X_n)$ is independent of the order of $\mathbf{x}^i$ in $\{\mathbf{x}^1, \cdots, \mathbf{x}^n\}$, and (2.9) also serves as an equivalent definition for $T(\mathbf{x}|X_n)$.

From (2.2) and (2.9), it is also obvious that there is a close relationship between the box spline $M(\cdot|X_n)$ and the multivariate truncated power $T(\cdot|X_n)$. The following result can be proved by mathematical induction.

THEOREM 2.9. *Let $X_n \subset \mathbf{Z}^s \backslash \{\mathbf{0}\}$ with $\langle X_n \rangle = \mathbf{R}^s$. Then*

(2.10) $$M(\mathbf{x}|X_n) = \triangle_{X_n} T(\mathbf{x}|X_n).$$

In the univariate setting, (2.10) becomes (1.4), namely:

$$B_n(x) = \frac{1}{(n-1)!} \triangle^n x_+^{n-1}.$$

To state the next result, we need the notion of "discrete truncated powers." Let

$$c(X_n) = \left\{\sum_{i=1}^n t_i \mathbf{x}^i: \ t_i \in \mathbf{Z}_+, i = 1, \cdots, n\right\}$$

which may be called the *discrete affine cone* defined by the direction set X_n. Recall that by replacing $\mathbf{x}^i$ with $-\mathbf{x}^i$, if necessary, the first nonzero component of each $\mathbf{x}^i$ is assumed to be positive. The *discrete truncated power* $t(\mathbf{k}|X_n)$ is defined as follows:

DEFINITION. Set

$$t(\mathbf{k}|\mathbf{x}^1,\cdots,\mathbf{x}^s) = \begin{cases} 1 & \text{if } \mathbf{k}\in c(X_s) \\ 0 & \text{otherwise}, \end{cases}$$

and define inductively, for $m = s+1,\cdots,n$,

$$t(\mathbf{k}|\mathbf{x}^1,\cdots,\mathbf{x}^m) = \sum_{j=0}^{\infty} t(\mathbf{k}-j\mathbf{x}^m|\mathbf{x}^1,\cdots,\mathbf{x}^{m-1}).$$

Then $t(\mathbf{k}|X_n) = t(\mathbf{k}|\mathbf{x}^1,\cdots,\mathbf{x}^n)$.

THEOREM 2.10.

$$(2.11) \qquad T(\mathbf{x}+\frac{1}{2}\sum_{i=1}^{n}\mathbf{x}^i|X_n) = \sum_{\mathbf{k}\in c(X_n)} t(\mathbf{k}|X_n)M(\mathbf{x}-\mathbf{k}|X_n).$$

This identity can be verified by using mathematical induction.

2.4. Box spline series. In the univariate case, every spline function of order n and with knots at $j+\frac{n}{2}$, $j=0,\pm 1,\cdots$, can be expressed as

$$\sum_{i\in\mathbf{Z}} c_i B_n(\cdot - i)$$

which is called a B-spline series. Consequently, the B-spline $B_n(x)$ itself determines the spline space. In fact, $\{B_n(\cdot - j) : j \in \mathbf{Z}\}$ is a linearly independent set. However, these properties cannot be generalized to the multivariate setting as we will see in the following as well as the next chapter. First, the "spline space" S_d^r on a given grid, with degree d and smoothness class C^r, usually contains more than one box spline, and in addition, the collection of all multi-integer translates of a single box spline may not be linearly independent. Further, it will be seen in the next chapter that some spline functions cannot be represented as linear combinations of translates of the locally supported ones. Let us first discuss linear independence of multi-integer translates of a single box spline.

Let $X_n = \{\mathbf{x}^1, \cdots, \mathbf{x}^n\} \subset \mathbf{Z}^s \backslash \{\mathbf{0}\}$ with $\langle X_n \rangle = \mathbf{R}^s$. For any subset $Y_s = \{\mathbf{x}^{i_1}, \cdots, \mathbf{x}^{i_s}\}$ of X_n, Y_s will also be used to denote the matrix whose jth column is $\mathbf{x}^{i_j}$.

THEOREM 2.11. *The collection* $\{M(\cdot - \mathbf{j}|X_n): \mathbf{j} \in \mathbf{Z}^s\}$ *is linearly independent if and only if*
$$|\det Y_s| = 1$$
for all $Y_s \subseteq X_n$ *with* $\langle Y_s \rangle = \mathbf{R}^s$.

We refer the reader to the original papers of de Boor and Höllig [29], Dahmen and Micchelli [98], and Jia [141] for the proof of this result.

Regardless of linear independence, let us consider the vector space $S(X_n)$ of the box spline series
$$\sum_{\mathbf{j} \in \mathbf{Z}^s} c_\mathbf{j} M(\cdot - \mathbf{j}|X_n).$$

To study how well this subspace can be used for approximation, it is important to know what polynomials lie in $S(X_n)$. Let π_d^s denote the space of all polynomials in $\mathbf{R}^s$ of total degree d and $\pi^s = \bigcup_d \pi_d^s$. The following can be found in de Boor and Höllig [29].

THEOREM 2.12.
$$\pi^s \cap S(X_n) = \bigcap_{\substack{Y \subset X_n \\ \langle X_n \backslash Y \rangle \neq \mathbf{R}^s}} \ker D_Y.$$

As a consequence of this result, it can be shown that if
$$r(X_n) = \min\{\#Y : Y \subset X_n, \langle X_n \backslash Y \rangle \neq \mathbf{R}^s\} - 2$$
as defined in (iv) of Theorem 2.2, then
$$\pi^s_{r(X_n)+1} \subset S(X_n)$$
and
$$\pi^s_{r(X_n)+2} \not\subset S(X_n).$$

Hence, as expected, the order of approximation from $S(X_n)$ is $r(X_n) + 2$. This is proved in de Boor and Höllig [29]. To be more precise, let us consider the scaling operator
$$(\sigma_h f)(\mathbf{x}) = f(\frac{1}{h}\mathbf{x}), \quad h > 0,$$

and set
$$S_h(X_n) = \{\sigma_h f : f \in S(X_n)\}.$$

THEOREM 2.13. *The approximation order of $S(X_n)$ is $r(X_n) + 2$. That is,*

(2.12) $$\text{dist}_{L^p}(f, S_h(X_n)) = O(h^{r(X_n)+2})$$

for any sufficiently smooth function $f(\mathbf{x})$ in $L^p(\mathbf{R}^s)$, $1 \le p \le \infty$, but there exists a C^∞ function $f_0 \in L^p(\mathbf{R}^s)$ such that

$$h^{-r(X_n)-2}\text{dist}_{L^p}(f_0, S_h(X_n)) \not\to 0.$$

The order of convergence in (2.12) can also be verified by using the Fourier transformation formula (2.3) and the S-F condition to be discussed in Chapter 8. Indeed, by setting

$$\rho(X_n) = \max_{\mathbf{j} \in \mathbf{Z}^s \setminus \{\mathbf{0}\}} \max\{\#W : \mathbf{j} \perp W \subset X_n\}$$

it follows that $r(X_n) + \rho(X_n) = n - 2$, and on the other hand, the order of the zero $2\pi\mathbf{j}$ of $\widehat{M}(\cdot|X_n)$, $\mathbf{j} \in \mathbf{Z}^s \setminus \{\mathbf{0}\}$, is $n - \#W$ where W is the maximal subset of X_n orthognal to $\mathbf{j}$. We remark, however, that the approximation order from the subspace generated by more than one box spline is still an open problem, although some important cases have been settled recently by de Boor and Höllig [32]. This will be discussed in §4.4. Explicit formulas which give the maximum order of approximation may be called quasi-interpolants. This subject will be discussed in Chapter 8.

CHAPTER 3

Bivariate Splines on Three- and Four-Directional Meshes

Let M and N be positive integers and consider the rectangular region $\Omega_{MN} = [0, M+1] \times [0, N+1]$. If the rectangular partition $x - i = 0$ and $y - j = 0$, $i, j \in \mathbf{Z}$, of Ω_{MN} is refined by drawing in all diagonals $x - y - k = 0$, $k \in \mathbf{Z}$, we obtain the so-called type-1 triangulation $\triangle_{MN}^{(1)}$ of Ω_{MN}. If it is further refined by drawing in the other diagonals $x + y - \ell = 0$, $\ell \in \mathbf{Z}$, we have the so-called type-2 triangulation $\triangle_{MN}^{(2)}$ of Ω_{MN}. Hence, the type-1 triangulation of Ω_{MN} gives a three-directional mesh, while the type-2 triangulation gives a four-directional mesh. More descriptive terminologies are unidiagonal and crisscross partitions, respectively. For any integers r and d, let $S_d^r(\triangle_{MN}^{(i)})$, $i = 1, 2$, denote the vector space of all functions in $C^r(\Omega_{MN})$ whose restrictions to each triangular cell of the grid partition $\triangle_{MN}^{(i)}$ of Ω_{MN} are in π_d^2, the collection of all polynomials in two variables of total degree at most d. This chapter is devoted to a brief discussion of the "bivariate spline" spaces $S_d^r(\triangle_{MN}^{(i)})$, with emphasis on dimension, locally supported splines and the span of their translates, minimal supports, basis, and approximation orders. Of course, these are fairly simple but very important properties of univariate splines which were discussed in Chapter 1.

3.1. Dimension. In order to determine how many "independent" locally supported functions there are in a certain spline space with "regular" grid partition, we must know its dimension, at least asymptotically. For the spaces $S_d^r(\triangle_{MN}^{(i)})$, the dimension is of the form $aMN + bM + cN + e$ for some integers a, b, c, and e, and the number of "independent" locally supported functions is the coefficient a, since the others can be written as finite linear combinations of translates of them. In Chui and Wang [73], the exact dimension of $S_d^r(\triangle_{MN}^{(i)})$ is determined. In fact, a more general result, which will be discussed in the next chapter, is obtained.

THEOREM 3.1. *Let $d, r \in \mathbf{Z}_+$. Then*

(3.1)
$$\dim S_d^r(\triangle_{MN}^{(1)}) = a_d^r MN + (d-r)_+(d-r+1)(M+N) \\ + \binom{d-r+1}{2} + \binom{d+2}{2}$$

and

(3.2)
$$\dim S_d^r(\triangle_{MN}^{(2)}) = b_d^r MN + \frac{3}{2}(d-r)_+(d-r+1)(M+N) + 2\binom{d-r+1}{2} + \binom{d+2}{2},$$

where

(3.3)
$$a_d^r = (d - 2r + [\frac{r+1}{2}])(d - r - [\frac{r+1}{2}])_+$$

and

(3.4) $\quad b_d^r = \frac{1}{2}(d - r - [\frac{r+1}{3}])_+(3d - 5r + 3[\frac{r+1}{3}] + 1) + \binom{d-2r}{2}.$

Here, $[x]$ denotes, as usual, the largest integer which does not exceed x, and the binomial coefficient $\binom{y}{2}$ is defined to be 0 if $y < 2$.

Hence, for the bivariate spline spaces on the three- and four-directional meshes, we expect to find a_d^r and b_d^r "independent" locally supported functions, respectively. In applications, however, we would like to use the smoothest splines with the lowest degree but, at the same time, be able to do the approximation. That is, we are interested in working with the spaces $S_d^r(\triangle_{MN}^{(i)})$ where, for a given $r \in \mathbf{Z}_+$, d is the smallest so that a_d^r or b_d^r are nonzero.

We will use the notation $d^* = d^*(r, i)$ for the smallest d such that $a_d^r > 0$ for $i = 1$ and $b_d^r > 0$ for $i = 2$ and denote by a_r^*, b_r^* the corresponding values of a_d^r, b_d^r.

From (3.3) and (3.4), we have the following table.

r	$2k-1$	$2k$	r	$3k$	$3k+1$	$3k+2$
d^*	$3k$	$3k+1$	d^*	$4k+1$	$4k+2$	$4k+4$
a_r^*	2	1	b_r^*	2	1	3
$S_d^r(\triangle_{MN}^{(1)})$			$S_d^r(\triangle_{MN}^{(2)})$			

That is, for the three-directional mesh, there are one or two "independent" locally supported splines with minimal degree, while for the four-directional mesh, there are up to three "independent" locally supported ones with minimal degree, depending on the smoothness requirement.

3.2. Locally supported splines.

We start with the following example.

Example 3.1. Recall the notation of box splines

$$M_{tuvw} \quad \text{and} \quad M_{tuv} := M_{tuv0}$$

discussed in Example 2.2. Set

$$(3.5) \quad B_{tuvw}(x,y) = M_{tuvw}(\mathbf{x} - \frac{t}{2}\mathbf{e}^1 - \frac{u}{2}\mathbf{e}^2 - \frac{v}{2}(\mathbf{e}^1 + \mathbf{e}^2) - \frac{w}{2}(\mathbf{e}^2 - \mathbf{e}^1))$$

where $\mathbf{x} = (x,y)$. It follows that

$$B_{tuvw} \in S_d^r(\triangle_{MN}^{(2)})$$

with $d = t + u + v + w - 2$ and $r = r(X_n)$ defined in (iv) of Theorem 2.2. Of course, if $w = 0$, we have

$$B_{tuv} \in S_d^r(\triangle_{MN}^{(1)}).$$

Note that for a given r, the minimal degree is attained when the integers t, u, v (in the three-directional mesh) or t, u, v, w (in the four-directional mesh) are "balanced." For instance, we have:

$$(3.6) \quad \begin{cases} B_{111} \in S_1^0(\triangle_{MN}^{(1)}); & B_{221}, B_{212}, B_{122} \in S_3^1(\triangle_{MN}^{(1)}); \\ B_{222} \in S_4^2(\triangle_{MN}^{(1)}); & B_{332}, B_{323}, B_{233} \in S_6^3(\triangle_{MN}^{(1)}) \end{cases}$$

and

$$(3.7) \quad \begin{cases} B_{1110}, B_{1101}, B_{1011}, B_{0111} \in S_1^0(\triangle_{MN}^{(2)}); \\ B_{1111} \in S_2^1(\triangle_{MN}^{(2)}); \quad B_{2220}, B_{2202}, B_{2022}, B_{0222}, \\ B_{2211}, B_{2121}, B_{2112}, B_{1221}, B_{1212}, B_{1122} \in S_4^2(\triangle_{MN}^{(2)}). \end{cases}$$

Let us investigate some of the examples of box splines in (3.6) and (3.7) more carefully.

Example 3.2. From (3.1), we know that

$$\dim S_1^0(\triangle_{MN}^{(1)}) = (M+2)(N+2),$$

which is the number of vertices in the triangulation $\triangle_{MN}^{(1)}$ of Ω_{MN}. Hence, it is easy to see that

$$\{B_{111}(x-i, y-j): \quad i = -1, \cdots, M; \ j = -1, \cdots, N\}$$

is a basis of $S_1^0(\triangle_{MN}^{(1)})$.

Example 3.3. From (3.1), we know that

$$\dim S_3^1(\triangle_{MN}^{(1)}) = 2MN + 6(M+N) + 13.$$

On the other hand, there are three box splines $B_{221}, B_{212}, B_{122}$ whose translates yield at least $3MN$ nontrivial functions in $S_3^1(\triangle_{MN}^{(1)})$. Hence, they must be linearly dependent. It is interesting to note that removing one of the three box splines does not yield a linearly independent set either. In addition, these box splines do *not* have minimal supports. We will discuss this issue in Example 3.8 in §3.4.

Example 3.4. There is only one box spline B_{222} in $S_4^2(\triangle_{MN}^{(1)})$ and the number of $B_{222}(x-i, y-j)$, where $i, j \in \mathbf{Z}$, which do not vanish identically on Ω_{MN} is $(M+4)(N+4) - 2$ or

$$MN + 4(M+N) + 14.$$

However, from (3.1), we have

$$\dim S_4^2(\triangle_{MN}^{(1)}) = MN + 6(M+N) + 18.$$

Hence, although $\{B_{222}(x-i, y-j)\}$ is linearly independent (see Theorem 2.11), it does not span $S_4^2(\triangle_{MN}^{(1)})$. In Chui and Wang [75], it is shown that a basis of $S_4^2(\triangle_{MN}^{(1)})$ is given by

$$\{B_{222}(x-i, y-j), (x-p)_+^4, (y-q)_+^4, (x-y-r)_+^4\}$$

where $p = 0, \cdots, M$; $q = 0, \cdots, N$; $r = -N-1, \cdots, M$; and all i, j such that $B_{222}(x-i, y-j)$ does not vanish identically on Ω_{MN}. We will see in §3.4 that globally supported elements cannot be avoided in giving a basis of some bivariate spline spaces.

Example 3.5. There are four box splines in $S_1^0(\triangle_{MN}^{(2)})$ but by (3.1) the dimension of this space is $2MN + 3(M+N) + 5$. So again we need only two of them. However, it is clear that none of them has minimal support. The two minimally supported ones, which are piecewise linear polynomials denoted by $f_2^0(x, y)$ and $f_3^0(x, y)$, can be defined by taking the value 1 at a vertex and zero at all the other vertices. In Figure 3.1, we indicate their supports.

supp f_2^0

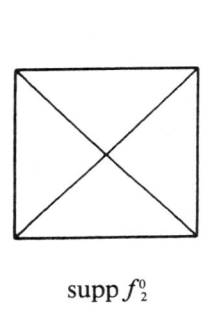

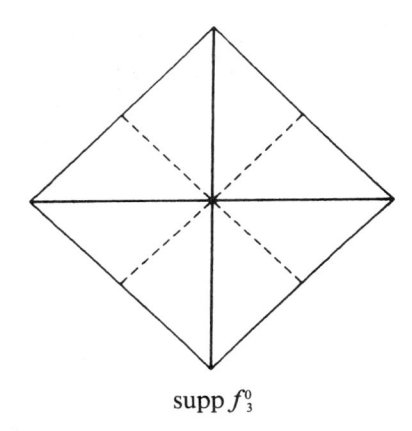

supp f_3^0

FIG. 3.1

Hence, the total number of such splines is the same as the number of vertices, namely:

$$(M+2)(N+2) + (M+1)(N+1)$$
$$= 2MN + 3(M+N) + 5$$

which is the same as the dimension. It is clear that they generate the whole space $S_1^0(\triangle_{MN}^{(2)})$, and thus, give a basis for this space.

Example 3.6. $S_2^1(\triangle_{MN}^{(2)})$ is a very interesting space. It has dimension $MN+3(M+N)+8$. However, there are $(M+3)(N+3) = MN+3(M+N)+9$ box splines $B_{1111}(x-i, y-j)$ that do not vanish identically on Ω_{MN}. Hence, they must be linearly dependent there. The dependence relationship

$$\sum_{i,j}(-1)^{i+j}B_{1111}(x-i, y-j) = 0$$

for all $(x,y) \in \Omega_{MN}$ is given in Chui and Wang [76], where it is also shown that if any of these functions is removed, the remaining functions form a basis of the space.

Example 3.7. The space $S_4^2(\triangle_{MN}^{(2)})$ is perhaps the most interesting. It has ten box splines according to (3.7), but since $b_2^* = 3$, only three are useful. However, none of them has minimal support. Sablonnière [179] gives two minimally supported splines f_4^0 and f_5^0 in $S_4^2(\triangle_{MN}^{(2)})$ which are displayed in the following figure, where the numbers shown are 48 times the Bézier nets and those not shown are either zero or can be easily obtained by symmetry. (We will discuss Bézier nets in Chapter 5.) In Chui and He [56] it is shown that these two are the only minimally supported splines in this space, at least among those with convex supports. One must be very careful in considering uniqueness in $\mathbf{R}^s$ for $s \geq 2$. Here, we mean that if g is a minimally supported spline in $S_4^2(\triangle_{MN}^{(2)})$ with convex support, then it must be a constant multiple of a translate of f_4^0 or f_5^0. A proof of this statement must allow supports of different convex shapes which may not even be symmetric. If the "convexity" condition is not imposed, the proof of uniqueness is still not known. Since there must be three "independent" locally supported splines in $S_4^2(\triangle_{MN}^{(2)})$, the notion of quasi-minimally supported splines is introduced in Chui and He [56]. We will give the definition in the next section.

A quasi-minimally supported $S_4^2(\triangle_{MN}^{(2)})$ spline, denoted by f_6^0, is displayed in Figure 3.2. Again the numbers shown are 48 times the Bézier net.

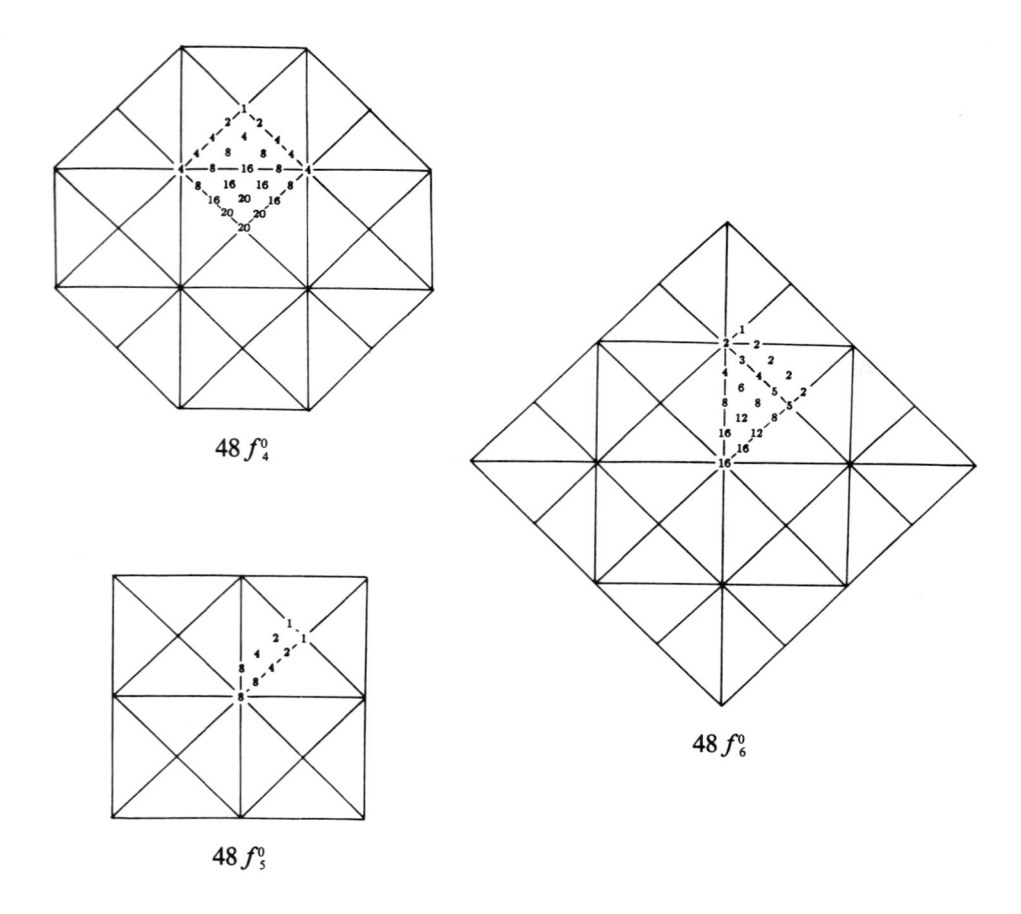

FIG. 3.2

3.3. Minimally and quasi-minimally supported bivariate splines.

A function in a certain multivariate spline space is said to have minimal support if there does not exist a nontrivial function in the same space that vanishes identically outside any proper subset of this support. A function in the multivariate spline space is said to have quasi-minimal support D if it is not in the linear span of all minimally supported functions in the space whose supports lie in D, but any function in this space whose support is properly contained in D must be in this linear span. It is well known that all univariate B-splines have minimal supports, but since they form a basis of the spline space there does not exist any quasi-minimally supported univariate spline function.

So far, only the three- and four-directional meshes in $\mathbf{R}^2$ have been considered in the literature, and the only results are for those spaces with smallest degrees. That is, the spaces that have been studied are:

$$(3.8) \qquad S_{3k}^{2k-1}(\triangle_{MN}^{(1)}), \quad S_{3k+1}^{2k}(\triangle_{MN}^{(1)})$$

and

$$(3.9) \qquad S_{4k}^{3k-1}(\triangle_{MN}^{(2)}), \quad S_{4k+1}^{3k}(\triangle_{MN}^{(2)}), \quad S_{4k+2}^{3k+1}(\triangle_{MN}^{(2)}),$$

where $k \in \mathbf{Z}_+$.

The case $k = 0$ in (3.8) is trivial. The minimally supported functions in $S_0^{-1}(\triangle_{MN}^{(1)})$ are the characteristic functions χ_A and χ_B where the sets A and B are shown in Figure 3.3. There is only one minimally supported spline in $S_1^0(\triangle_{MN}^{(1)})$, and it is the box spline $B_{111}(x, y)$ as discussed in Examples 2.2 and 3.2.

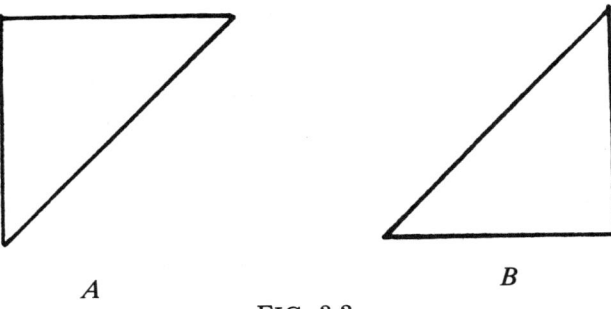

FIG. 3.3

For the four-directional mesh, it turns out that it is *not* appropriate to consider $S_0^{-1}(\triangle_{MN}^{(2)})$ since it would be difficult to describe the general pattern for higher degree splines in $S_{4k}^{3k-1}(\triangle_{MN}^{(2)})$. The two minimally supported splines in $S_1^0(\triangle_{MN}^{(2)})$ have been discussed in Example 3.5, and were called $f_2^0(x,y)$ and $f_3^0(x,y)$ there. The box spline $B_{1111}(x,y)$ is a minimum supported spline, and in fact the only one in $S_2^1(\triangle_{MN}^{(2)})$ (see Example 3.6). We denote $f_1^0 = B_{1111}$. To describe the spaces $S_{4k}^{3k-1}(\triangle_{MN}^{(2)})$ in (3.9), it is better to start with $k=1$, namely the space $S_4^2(\triangle_{MN}^{(2)})$ as discussed in Example 3.7. The two minimally supported splines in this space are f_4^0 and f_5^0 constructed in Sablonnière [179] and the quasi-minimally supported one in this space is f_6^0 constructed in Chui, He, and Wang [62]. That f_6^0 is quasi-minimally supported and is unique is proved in Chui and He [56]. Again, when we consider uniqueness, we only restrict ourselves to those with convex supports but otherwise having arbitrary shapes. Although we believe that the "convexity" condition can be dropped, there is still no proof available.

To unify our notation, the initial functions in the three-directional mesh will be denoted by:

(3.10) $$g_1^0 = B_{111}, \quad g_2^0 = \chi_A, \quad g_3^0 = \chi_B$$

and the ones in the four-directional mesh by

(3.11) $$f_1^0 = B_{1111}, \quad f_2^0, \quad f_3^0, \quad f_4^0, \quad f_5^0, \quad f_6^0$$

discussed earlier. We now set

(3.12) $$g_i^k = \underbrace{g_1^0 * \cdots * g_1^0}_{k} * g_i^0, \qquad i=1,2,3$$

and

(3.13) $$f_i^k = \underbrace{f_1^0 * \cdots * f_1^0}_{k} * f_i^0 \quad i=1,\cdots,6.$$

Note that in (3.12) the convolution is taken with the linear box spline $g_1^0 = B_{111}$, and in (3.13) we convolve with the quadratic box spline $f_1^0 = B_{1111}$, to yield g_i^k and f_i^k, respectively.

It is not difficult to show that

(3.14) $$g_1^k \in S_{3k+1}^{2k}(\triangle_{MN}^{(1)}) \quad ; \quad g_2^k, g_3^k \in S_{3k}^{2k-1}(\triangle_{MN}^{(1)})$$

and

(3.15) $$f_1^k \in S_{4k+2}^{3k+1}(\triangle_{MN}^{(2)}); \quad f_2^k, f_3^k \in S_{4k+1}^{3k}(\triangle_{MN}^{(2)});$$
$$f_4^k, f_5^k, f_6^k \in S_{4k+4}^{3k+2}(\triangle_{MN}^{(2)});$$

for $k = 0, 1, \cdots$. It should be noted that among all the functions in (3.14) and (3.15), only g_1^k and f_1^k are box splines.

Recall from Chapter 2 that a box spline is obtained by taking convolution along a set of directions starting from the characteristic function of an affine cube. The functions in (3.14) and (3.15) are obtained by taking convolution with the "initial" functions g_i^0 and f_i^0 along each of the three directions for g_i^k, and each of the four directions for f_i^k, evenly.

THEOREM 3.2. *For all* $k \in \mathbf{Z}_+$, $g_1^k, g_2^k, g_3^k, f_1^k, f_2^k, f_3^k, f_4^k, f_5^k$ *are minimally supported splines in the corresponding spaces as indicated in* (3.14) *and* (3.15). *For each* $k = 0, 1, 2, \cdots$, f_6^k *is a quasi-minimally supported spline in* $S_{4k+4}^{3k+2}(\triangle_{MN}^{(2)})$. *Furthermore, all the above minimally supported splines are unique in the sense that any minimally supported spline with convex support in one of the spaces must be a constant multiple of a translate of one of them in the same space, and* f_6^k *is unique in the sense that if* f *is a quasi-minimally supported spline with convex support in* $S_{4k+4}^{3k+2}(\triangle_{MN}^{(2)})$ *then*

$$f(\cdot) = c f_6^k(\cdot - \mathbf{j}) + d f_5^k(\cdot - \mathbf{j})$$

for some constants c *and* d, *and some* $\mathbf{j} \in \mathbf{Z}^2$.

Note that the support of f_4^k is not contained in supp f_6^k. The first statement in the above theorem is proved in de Boor and Höllig [33], and the proof of the other two statements can be found in Chui and He [56].

To facilitate computation and simplify bivariate convolution, we define:

$$(I_1 f)(x, y) = \int_{x-1}^{x} f(s, y) ds$$

$$(I_2 f)(x, y) = \int_{y-1}^{y} f(x, t) dt$$

$$(I_3 f)(x, y) = \int_{x-1}^{x} f(u, u - x + y) du$$

$$(I_4 f)(x, y) = \int_{x-1}^{x} f(v, -v + x + y) dv.$$

Then it can be shown that for $k = 0, 1, \cdots$, we have

(3.16) $$g_i^{k+1} = I_3 I_2 I_1 g_i^k \quad , \quad i = 1, \cdots, 3,$$

and

(3.17) $$f_i^{k+1} = I_4 I_3 I_2 I_1 f_i^k \quad , \quad i = 1, \cdots, 6.$$

These formulas will be used to give the Bézier representations of the bivariate splines g_i^k and f_i^k in Chapter 7. It is clear from the definition that the g_i^k's and f_i^k's in the same spaces together give a partition of unity, namely:

(3.18)
$$\begin{cases} \sum_{\mathbf{j} \in \mathbf{Z}^2} g_1^k(\cdot - \mathbf{j}) \equiv 1 \\ \sum_{\mathbf{j} \in \mathbf{Z}^2} (g_2^k(\cdot - \mathbf{j}) + g_3^k(\cdot - \mathbf{j})) \equiv 1 \\ \sum_{\mathbf{j} \in \mathbf{Z}^2} f_1^k(\cdot - \mathbf{j}) \equiv 1 \\ \sum_{\mathbf{j} \in \mathbf{Z}^2} (f_2^k(\cdot - \mathbf{j}) + f_3^k(\cdot - \mathbf{j})) \equiv 1 \\ \sum_{\mathbf{j} \in \mathbf{Z}^2} (f_4^k(\cdot - \mathbf{j}) + f_5^k(\cdot - \mathbf{j}) + f_6^k(\cdot - \mathbf{j})) \equiv 1 \,. \end{cases}$$

3.4. Bases and approximation order. The B-splines in the univariate case give a basis of the spline space, and the approximation order from this space is the same as the order of the splines. More precisely, if $\mathcal{S}_{\mathbf{t},n}$ is a univariate spline space of order n (or degree $n-1$) and with knot sequence $\mathbf{t}$, then the collection of B-splines $\{N_{\mathbf{t},n,i}\}$ is a basis of $\mathcal{S}_{\mathbf{t},n}$, and for any $f \in C^n$, we have

$$\text{dist}_{L^\infty}(f, \mathcal{S}_{\mathbf{t},n}) \leq C_f h^n$$

for some constant C_f depending only on the function f, where

$$h = \max(t_{i+1} - t_i)$$

but on the other hand, there is a function g in C^∞ with

$$\liminf_{h \to 0} h^{-n} \text{dist}_{L^\infty}(g, \mathcal{S}_{\mathbf{t},n}) > 0 \,,$$

(see § 1.4). These univariate results do not have direct generalizations to the multivariate setting, even on the three- and four-directional meshes in $\mathbf{R}^2$. We begin with two examples studied in Chui and Wang [75].

supp g_2^1 supp g_3^1

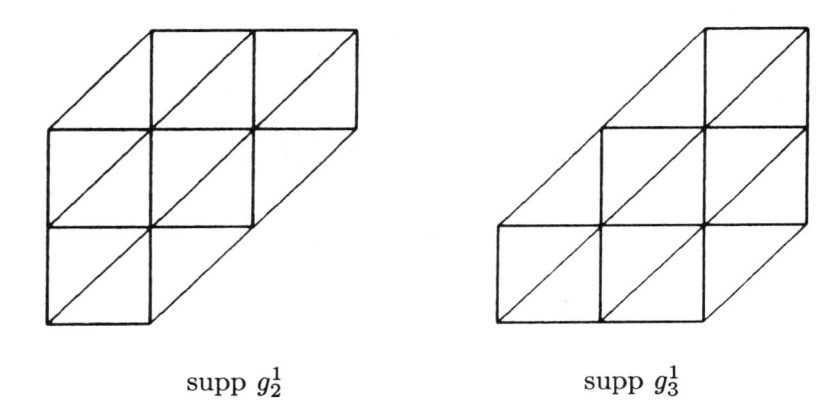

FIG. 3.4

Example 3.8. Consider the space $S_3^1(\triangle_{MN}^{(1)})$. By Theorem 3.2, it has two minimally supported splines g_2^1 and g_3^1 whose supports are given in Figure 3.4. These functions are constructed in Fredrickson [113] and their supports are smaller than those of the box splines $B_{221}, B_{212}, B_{122}$. It turns out, however, that the total number of translates of g_2^1 and g_3^1 that do not vanish identically on Ω_{MN} is three larger than the dimension of $S_3^1(\triangle_{MN}^{(1)})$, and hence, these translates must be linearly dependent on Ω_{MN}. In fact, the dependence relationship is given by:

(i) $\sum_{i,j}[g_2^1(x-i, y-j) - g_3^1(x-i, y-j)] = 0$,

(ii) $\sum_{i,j}[(i+\tfrac{1}{3})g_2^1(x-i, y-j) - (i-\tfrac{1}{3})g_3^1(x-i, y-j)] = 0$,

and

(iii) $\sum_{i,j}[(j-\tfrac{1}{3})g_2^1(x-i, y-j) - (j+\tfrac{1}{3})g_3^1(x-i, y-j)] = 0$

for all $(x,y) \in \Omega_{MN}$. Certain geometric conditions on the location of the centers of the supports of $g_2^1(x-i, y-j)$ and $g_3^1(x-i, y-j)$ can be described that guarantee which three functions are to be deleted to give a minimally supported basis (cf. [75]).

Example 3.9. The space $S_4^2(\triangle_{MN}^{(1)})$ is an interesting and "well-behaved" one. It has only one box spline B_{222} which also has minimal support, being $g_1^1 = g_1^0 * g_1^0 = B_{111} * B_{111}$. However, the difference between dim $S_4^2(\triangle_{MN}^{(1)})$ and the number of translates of B_{222} that do not vanish identically on Ω_{MN} is $2(M+N)+4$, which is the same as the cardinality of the "crosscut" functions:

(3.19) $$\{(x-i)_+^4, (y-j)_+^4, (x-y-r)_+^4: \quad i=0,\cdots,M; \\ j=0,\cdots,N; \; r=-N-1,\cdots,M\}.$$

Indeed, as we mentioned in Example 3.5, this set together with the translates of B_{222} give a basis of the space $S_4^2(\triangle_{MN}^{(1)})$ (cf. [75]). Note that we call the functions in (3.19) *crosscut functions* in order not to confuse the reader with the *truncated powers* introduced by Dahmen [79] (cf. [82, §2.3]), since truncated powers in the multivariate setting can be generated by translates of box splines in the same space, (cf. (2.11) in §2.3 and for more details, see §4.2).

From the above two examples, it would seem quite difficult to produce a basis for $S_d^r(\triangle_{MN}^{(i)})$, $i=1,2$, with the maximum number of minimally supported elements. Nonetheless, Bamberger [8] gives very satisfactory results in this direction. We will only mention Bamberger's generalization of Example 3.9.

THEOREM 3.3. *Let $r \geq 2$. The collection*

$$\{B_{rrr}(x-i,y-j), x^a y^b (x-k)_+^{2r}, x^a y^b (y-\ell)_+^{2r}, x^a y^b (x-y-m)_+^{2r}: \\ a+b \leq r-2, k=0,\cdots,M, \ell=0,\cdots,N, \\ m=-N-1,\cdots,M, \\ B_{rrr}(x-i,y-j) \text{ nontrivial on } \Omega_{MN}\}$$

is a basis of $S_{3r-2}^{2r-2}(\triangle_{MN}^{(1)})$.

In general, since linear independence is difficult to determine and is sometimes not a very important issue, we will only worry about the spanning property. We will, however, be concerned with the maximum number of minimally and quasi-minimally supported functions in the spanning set.

For each $r \in \mathbf{Z}_+$ let $d^* = d^*(r,i)$ be the smallest integer d such that $S_d^r(\triangle_{MN}^{(i)})$ has at least one locally supported function (cf. §3.1). We will only consider the spaces $S_{d^*}^r(\triangle_{MN}^{(i)})$. Let S denote the span of the translates of all the box splines in $S_{d^*}^r(\triangle_{MN}^{(i)})$ and T the spanning set of all the *generalized truncated powers*; that is, the collection of all crosscut functions and truncated powers (for more details, see §4.2). Then it is proved in de Boor and Höllig [31] and Dahmen and Micchelli [94] that

$$S_{d^*}^r(\triangle_{MN}^{(i)}) = S + T + \pi_{d^*}^2.$$

An important question is for which space $S^r_{d^*}(\triangle^{(i)}_{MN})$ does the collection of all minimally and quasi-minimally supported bivariate splines provide a basis for the space. In Chui and He [57] it is shown that

(i) for $S^r_{d^*}(\triangle^{(1)}_{MN})$,

$$\pi^2_{d^*}, T \subset S \iff d^* > \frac{3}{2}r,$$

and

(ii) for $S^r_{d^*}(\triangle^{(2)}_{MN})$,

$$\pi^2_{d^*}, T \subset S \iff d^* > \frac{4}{3}r.$$

Consequently, the following result is obtained.

THEOREM 3.4.
(i) Span $\{g^k_1(\cdot - \mathbf{j})\} = S^{2k}_{3k+1}(\triangle^{(1)}_{MN}) \iff k = 0$.
(ii) Span $\{g^k_2(\cdot - \mathbf{j}), g^k_3(\cdot - \mathbf{t})\} = S^{2k-1}_{3k}(\triangle^{(1)}_{MN}) \iff k = 0, 1$.
(iii) Span $\{f^k_1(\cdot - \mathbf{j})\} = S^{3k+1}_{4k+2}(\triangle^{(2)}_{MN}) \iff k = 0$.
(iv) Span $\{f^k_2(\cdot - \mathbf{j}), f^k_3(\cdot - \mathbf{t})\} = S^{3k}_{4k+1}(\triangle^{(2)}_{MN}) \iff k = 0, 1$.
(v) Span $\{f^k_4(\cdot - \mathbf{j}), f^k_5(\cdot - \mathbf{t}), f^k_6(\cdot - \mathbf{u})\} = S^{3k+2}_{4k+4}(\triangle^{(2)}_{MN}) \iff k = 0, 1$.

For the cases (i) – (iv), we can be more specific. In fact, Bamberger [8] gives a detail report on the bases for these cases.

We now turn to the study of approximation order. In §2.4, we have discussed this problem when only one box spline is used. Theorem 2.13 says that the approximation order of $S(X_n)$ is

$$r + 2$$

where $r = r(X_n)$, defined in (iv) of Theorem 2.2, indicates the order of smoothness of the box spline.

We now specialize in the bivariate spline spaces $S^r_d(\triangle^{(i)}_{MN}), i = 1, 2$, and study the order of approximation from these spaces, or more precisely, the spaces obtained by scaling the spline functions in $S^r_d(\triangle^{(i)}_{MN})$ by $h > 0$ (cf. §2.4). Of course, we are only interested in the situations where

$$d \geq d^*(r, i)$$

since we need locally supported splines for the purpose of approximation. It is interesting to note that if the smallest degree $d^* = d^*(r, i)$ is used, the approximation order is only $r + 2$, but not better, even if there might exist more than one minimal (or quasi-minimal) supported spline function. This result is obtained in Bamberger [8].

THEOREM 3.5. *The approximation orders of the bivariate spline spaces* $S^{2k}_{3k+1}(\triangle^{(1)}_{MN})$, $S^{2k-1}_{3k}(\triangle^{(1)}_{MN})$, $S^{3k+1}_{4k+2}(\triangle^{(2)}_{MN})$, $S^{3k}_{4k+1}(\triangle^{(2)}_{MN})$, *and* $S^{3k+2}_{4k+4}(\triangle^{(2)}_{MN})$ *are* $2k+2, 2k+1, 3k+3, 3k+2$, *and* $3k+4$, *respectively, for* $k = 0, 1, \cdots$.

Of course, when the degree d is allowed to be larger than $d^*(r,i)$, the approximation order may increase. To facilitate our discussion, we let

$$m_{i+2} = m_{i+2}(r,d)$$

denote the approximation order of the space $S^r_d(\triangle^{(i)}_{MN}), i = 1, 2$. Note that m_3 now denotes the approximation order in the discussion of the three-directional mesh, and m_4 denotes the four-directional mesh. Let

(3.20) $$M_3(r,d) = \min\{2(d-r), \quad d+1\}$$

and

(3.21) $$M_4(r,d) = \min\{3(d-r), \quad d+1\}.$$

We can now state the following result on approximation order from $S^r_d(\triangle^{(i)}_{MN}), i = 1, 2$.

THEOREM 3.6.

(3.22) $$\max(r+2, \quad M_3(r,d) - 2) \leq m_3 \leq M_3(r,d),$$

and

(3.23) $$r + 2 \leq m_4 \leq M_4(r,d).$$

The upper bound in (3.22) is obtained in de Boor and Höllig [31] where the conjecture that the lower bound should be

$$\max(r+2, \quad M_3(r,d) - 1)$$

is made, and the lower bound in (3.22) is obtained in Jia [143]. The upper bound in (3.23) is a result in Dahmen and Micchelli [94]. It would seem that there should be a better lower bound in (3.23), but we are not aware of its existence.

CHAPTER 4

Bivariate Spline Spaces

In the previous chapter we gave a brief description of bivariate spline spaces on the three- and four-directional meshes. We are now going to discuss the more general setting and study some of the interesting subspaces. In particular, we will give the dimension of a bivariate spline space and its super spline subspace on a quasi-crosscut partition. Lower and upper bounds of the dimension of spline and super spline spaces on an arbitrary triangulation are also established, and approximation order is discussed.

4.1. A classical approach. Since a univariate spline is a piecewise polynomial separated by points, it is natural to consider a bivariate spline as a piecewise polynomial separated by curves. We start with the following classical result of Bezout (cf. Walker [200]).

THEOREM 4.1. *If the number of common zeros of two real-valued polynomials in two variables is greater than the product of their (total) degrees, then these two polynomials must have a nontrivial common factor.*

As an interesting and important consequence, we have the following corollary.

COROLLARY 4.1. *Let $\ell(x,y)$ be an irreducible polynomial that defines an algebraic curve Γ; that is, $\Gamma: \ell(x,y) = 0$. Then its gradient $\nabla \ell(x,y)$ has at most a finite number of zeros that lie on Γ.*

Indeed, if the contrary holds, then $\ell(x,y)$ has nontrivial common factors with both its first partial derivatives, which have lower degrees. Since $\ell(x,y)$ is irreducible, the two partial derivatives must be identically zero, or $\ell(x,y)$ is a constant which cannot define a curve.

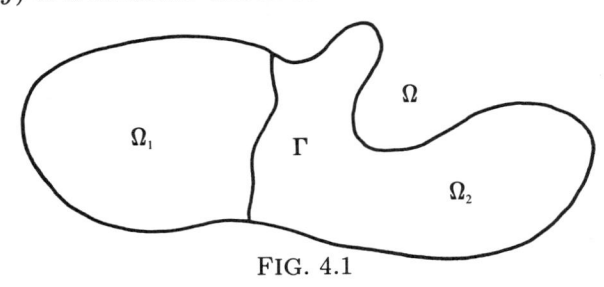

FIG. 4.1

Now suppose we have a region Ω which is divided into two subregions Ω_1 and Ω_2 by a curve Γ: $\ell(x,y) = 0$ (cf. Figure 4.1 above).

If $f \in C(\Omega)$ and $f\big|_{\Omega_1} = p_1$, $f\big|_{\Omega_2} = p_2$ are polynomials, then Γ must necessarily be an algebraic curve. Indeed, we have

$$\Gamma \subset \{(x,y): (p_1 - p_2)(x,y) = 0\}.$$

In the following result which can be found in Wang [201], we will assume that $\ell(x,y)$ is an irreducible polynomial.

THEOREM 4.2. *Let $p_1(x,y)$ and $p_2(x,y)$ be polynomials and $f\big|_{\Omega_1} = p_1$, $f\big|_{\Omega_2} = p_2$. If $f \in C^r(\Omega)$, then there is some polynomial $q(x,y)$ such that*

(4.1) $$p_2(x,y) = p_1(x,y) + [\ell(x,y)]^{r+1} q(x,y)$$

for all (x,y).

The reader might be interested in noting that (4.1) is in fact a generalization of (1.5) in the study of univariate splines. The proof of the above theorem is a simple application of Theorem 4.1 and Corollary 4.1 (cf. Chui and Wang [74]).

Now suppose that we have N adjacent regions $\Omega_1, \cdots, \Omega_N$ with a common grid point A and separated by algebraic curves $\Gamma_1, \cdots, \Gamma_N$, respectively, where Γ_i: $\ell_i(x,y) = 0$, as shown in Figure 4.2. Here, ℓ_i is an irreducible polynomial. For convenience, we set $\Gamma_{N+1} = \Gamma_1$. Let $\Omega = \bigcup_{i=1}^{N} \Omega_i$ and $f \in C^r(\Omega)$. Suppose that the restriction of $f(x,y)$ to each Ω_i is a polynomial $p_i \in \pi_d^2$. From Theorem 4.2, there exist polynomials $q_{1,2}, \cdots, q_{N,N+1}$ such that

(4.2) $$p_{i+1}(x,y) = p_i(x,y) + [\ell_i(x,y)]^{r+1} q_{i,i+1}(x,y)$$

where $i = 1, \cdots, N$ and $p_{N+1}(x,y) = p_1(x,y)$. Combining all the equations in (4.2), we have

(4.3) $$\sum_{i=1}^{N} q_{i,i+1}(x,y) [\ell_i(x,y)]^{r+1} = 0$$

for all (x,y). This is called the *conformality condition* at the vertex A. Solving this equation for the so-called *smoothing cofactors* $q_{i,i+1}(x,y)$, we have complete knowledge of $f(x,y)$ by using (4.2), once we know one of the N polynomials $p_1(x,y),\cdots,p_N(x,y)$, (cf. [74]). Let us consider the following example of characterizing bivariate spline spaces on a rectangular partition.

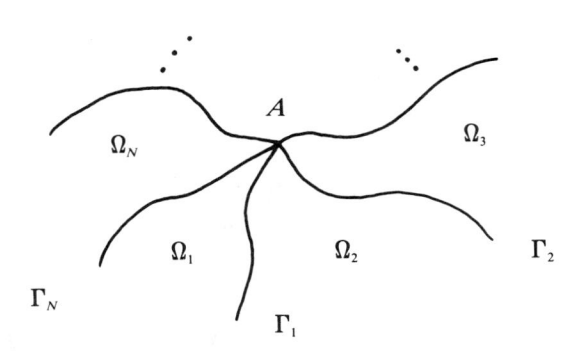

FIG. 4.2

Example 4.1. Let $\Omega = [a,b] \times [c,d]$ and $\Omega_{ij} = [x_i, x_{i+1}] \times [y_j, y_{j+1}]$, $i = 0,\cdots,M$, $j = 0,\cdots,N$, where $a = x_0 < \cdots < x_{M+1} = b$ and $c = y_0 < \cdots < y_{N+1} = d$. We are interested in studying the space S_d^r, $0 \le r < d < \infty$, of bivariate spline functions $f \in C^r(\Omega)$ with $f\Big|_{\Omega_{ij}} = p_{ij} \in \pi_d^2$.

By (4.1), we have

$$p_{ij}(x,y) = p_{i-1,j}(x,y) + (x - x_i)^{r+1} t_{ij}(x,y)$$

and

$$p_{ij}(x,y) = p_{i,j-1}(x,y) + (y - y_j)^{r+1} u_{ij}(x,y)$$

where $t_{i,j}, u_{ij} \in \pi_{d-r-1}^2$. Hence, for any (x,y) in Ω, say $(x,y) \in \Omega_{\ell m}$, we have:

$$\begin{aligned}
(4.4) \quad f(x,y) &= p_{\ell m}(x,y) \\
&= p_{\ell 0}(x,y) + \sum_{j=1}^{m}(y-y_j)^{r+1} u_{\ell j}(x,y) \\
&= p_{00}(x,y) + \sum_{i=1}^{\ell}(x-x_i)^{r+1} t_{i0}(x,y) \\
&\quad + \sum_{j=1}^{m}(y-y_j)^{r+1} u_{\ell j}(x,y).
\end{aligned}$$

This is, of course, a bad representation of $f(x,y)$ since one must know the location of (x,y). By using the notation of univariate truncated powers (cf. §1.1), we may rewrite (4.4) as

(4.5)
$$f(x,y) = p_{00}(x,y) + \sum_{i=1}^{M}(x-x_i)_+^{r+1} t_{i0}(x,y)$$
$$+ \sum_{j=1}^{N}(y-y_j)_+^{r+1} u_{\ell j}(x,y)$$

and the only location dependent notation left in (4.5) is $u_{\ell j}(x,y)$. We now appeal to the conformality condition (4.3) to remove the subscript ℓ, namely:
$$(x-x_\ell)^{r+1}(t_{\ell,j-1}(x,y) - t_{\ell j}(x,y))$$
$$+ (y-y_j)^{r+1}(u_{\ell j}(x,y) - u_{\ell-1,j}(x,y)) = 0.$$

Since $(x-x_\ell)^{r+1}$ and $(y-y_j)^{r+1}$ are relatively prime, we have $u_{\ell j}(x,y) - u_{\ell-1,j}(x,y) = (x-x_\ell)^{r+1} v_{\ell j}(x,y)$ for some $v_{\ell j} \in \pi^2_{d-2r-2}$. Hence, we may write

$$u_{\ell j}(x,y) = u_{0j}(x,y) + \sum_{k=1}^{\ell}(x-x_k)^{r+1} v_{kj}(x,y)$$
$$= u_{0j}(x,y) + \sum_{k=1}^{M}(x-x_k)_+^{r+1} v_{kj}(x,y)$$

and substituting this into (4.5) yields:

$$f(x,y) = p_{00}(x,y) + \sum_{i=1}^{M}(x-x_i)_+^{r+1} t_{i0}(x,y)$$
$$+ \sum_{j=1}^{N}(y-y_j)_+^{r+1} u_{0j}(x,y)$$
$$+ \sum_{j=1}^{N}\sum_{k=1}^{M}(x-x_k)_+^{r+1}(y-y_j)_+^{r+1} v_{kj}(x,y).$$

That is, a basis of the space S_d^r is given by

$$\{x^i y^j, x^k y^\ell (x-x_m)_+^{r+1}, x^k y^\ell (y-y_n)_+^{r+1}, x^p y^q (x-x_m)_+^{r+1}(y-y_n)_+^{r+1}\}$$

where $0 \leq i+j \leq d$, $0 \leq k+\ell \leq d-r-1$, $0 \leq p+q \leq d-2r-2$, and $m = 1, \cdots, M, n = 1, \cdots, N$. Of course, if $d < 2r + 2$, then this basis reduces to

$$\{x^i y^j, x^k y^\ell (x - x_m)_+^{r+1}, x^k y^\ell (y - y_n)_+^{r+1}\}$$

and the space S_d^r with $d < 2(r + 1)$ is not suitable for approximation purposes since its closure, as the mesh is refined, is the set of functions of the form

$$p(x, y) + p_1(x, y)f(x) + p_2(x, y)g(y)$$

where $p \in \pi_d^2$ and $p_1, p_2 \in \pi_{d-r-1}^2$.

4.2. Quasi-crosscut partitions. Let Ω be a simply connected region in $\mathbf{R}^2$ and partition Ω into subregions by drawing in straight line segments, each of which extends either from a boundary point to another boundary point of Ω, or from an interior grid point (also called an interior vertex) to the boundary of Ω. These lines are called crosscuts and rays, respectively. The partition obtained in this manner will be called a quasi-crosscut partition. For instance, the three- and four-directional meshes $\triangle_{MN}^{(1)}$ and $\triangle_{MN}^{(2)}$ considered in the previous chapter are quasi-crosscut partitions which consist only of crosscuts. The following dimension result can be found in Chui and Wang [73].

THEOREM 4.3. *Let Ω be a simply connected region in $\mathbf{R}^2$ and $\triangle$ a quasi-crosscut partition of Ω consisting of L crosscuts and a finite number of rays. Let $A_1, \cdots, A_V$ be the interior vertices of this partition and denote by N_i the total number of crosscuts or rays passing through or starting at A_i. Let $S_d^r(\triangle) = S_d^r(\triangle, \Omega)$ be the bivariate spline space of all functions $f(\mathbf{x})$ in $C^r(\Omega)$ such that the restrictions of $f(\mathbf{x})$ in each subregion of Ω under the partition $\triangle$ are polynomials in π_d^2. Then*

$$(4.6) \qquad \dim S_d^r(\triangle) = \binom{d+2}{2} + L \binom{d-r+1}{2} + \sum_{i=1}^{V} C_d^r(N_i)$$

where

$$(4.7) \qquad C_d^r(n) = \frac{1}{2}\left(d - r - \left[\frac{r+1}{n-1}\right]\right)_+ \left((n-1)d - (n+1)r\right.$$

$$\left. + (n-3) + (n-1)\left[\frac{r+1}{n-1}\right]\right)$$

and $[x]$ denotes, as usual, the largest integer not exceeding x.

We remark that the special case where $V = 1$ is given in Schumaker [185]. For bivariate C^1 quadratic splines observe that (4.6) reduces to

$$\dim S_2^1(\Delta) = 6 + L + \sum_{i=1}^{V}(N_i - 3)_+ .$$

If Δ is a simple quasi-crosscut partition; that is, $N_i = 2$ for all i, then

$$\dim S_d^r(\Delta) = \binom{d+2}{2} + L\binom{d-r+1}{2} + V\binom{d-2r}{2}$$

where $\binom{x}{2} = 0$ for $x < 2$, and this is consistent with Example 4.1. We also remark that the idea of the proof of Theorem 4.2 in Chui and Wang [73] is valid for the general case (when crosscuts and rays are replaced by algebraic curves that extend either from boundary to boundary or grid point to boundary) except that the value of the corresponding $C_d^r(n)$ has to be determined. There has been some recent interest in this study. We refer the reader to Billera [15], Stiller [193], and Whiteley [205]. For $n = 2$, when each vertex is shared by only two algebraic curves, the generalization is trivial. To demonstrate the idea of the proof of Theorem 4.3, let us consider the following example.

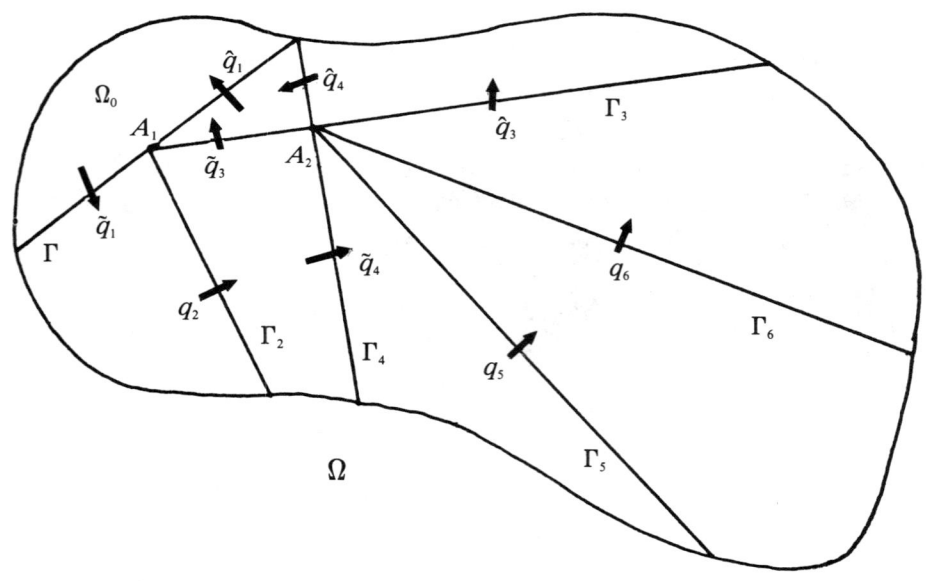

FIG. 4.3

Example 4.2. Consider the quasi-crosscut partition Δ of Ω in Figure 4.3 with two crosscuts, Γ_1 and Γ_4, and two (interior) vertices, A_1 and A_2, where $N_1 = 3$ and $N_2 = 4$. For any $f(x, y)$ in $S_d^r(\Delta, \Omega)$, let $p_0(x, y)$ in π_d^2 denote its restriction on some cell Ω_0, say, and to represent the function in the other cells, we simply use smoothing cofactors $\tilde{q}_1, \hat{q}_1, q_2, \tilde{q}_3, \hat{q}_3, \tilde{q}_4, \hat{q}_4, q_5,$ and q_6 as shown in Figure 4.3. These functions are polynomials in π_{d-r-1}^2 governed by the following conformality conditions at A_1 and A_2, respectively:

$$(\tilde{q}_1 + \hat{q}_1) \ell_1^{r+1} + q_2 \ell_2^{r+1} + \tilde{q}_3 \ell_3^{r+1} = 0$$

and

$$(-\tilde{q}_3 + \hat{q}_3) \ell_3^{r+1} + (\tilde{q}_4 + \hat{q}_4) \ell_4^{r+1} + q_5 \ell_5^{r+1} + q_6 \ell_6^{r+1} = 0,$$

which hold everywhere in $\mathbf{R}^2$. Here, $\ell_i \in \pi_1^2$ and

$$\Gamma_i = \{(x, y) : \ell_i(x, y) = 0\}, \quad i = 1, \cdots, 6.$$

The importance in the property of a quasi-crosscut partition is that the identities governed by the conformality conditions at all the interior vertices can always be uncoupled. Returning to our example, we note that although $\tilde{q}_3$ appears in both of the above governing identities, we may first solve for $\tilde{q}_3$ in the first one, and then $\hat{q}_3$ in the second one. In other words, for fixed $\tilde{q}_3$ and $\hat{q}_4$ in the second identity, by setting $q_3 = -\tilde{q}_3 + \hat{q}_3$ and $q_4 = \tilde{q}_4 + \hat{q}_4$, we note that q_3 and q_4 are completely free, since $\hat{q}_3$ and $\tilde{q}_4$ are. That is, the number of free parameters corresponding to the second identity, for fixed $\tilde{q}_3$ and $\hat{q}_4$, is the dimension of the solution space of (q_3, q_4, q_5, q_6) with

$$\sum_{i=3}^{6} q_i \ell_i^{r+1} = 0, \quad q_3, \cdots, q_6 \in \pi_{d-r-1}^2.$$

This dimension can be shown to be $C_d^r(4)$, (cf. [73],[185]). Similarly, for fixed $\hat{q}_1$ in π_{d-r-1}^2, the number of free parameters corresponding to the first identity is the dimension of the solution space of (q_1, q_2, q_3) with

$$\sum_{i=1}^{3} q_i \ell_i^{r+1} = 0, \quad q_1, \cdots, q_3 \in \pi_{d-r-1}^2,$$

where $q_1 = \tilde{q}_1 + \hat{q}_1$ and $q_3 = \tilde{q}_3$. Note that again $\tilde{q}_1$, and consequently q_1, is completely free although $\hat{q}_1$ is fixed. The dimension of this solution space is $C_d^r(3)$ (cf. [73],[185]). Finally, each of $\hat{q}_1$ and $\hat{q}_4$, which were fixed in the above argument, contributes $\binom{d-r+1}{2}$ parameters. Note that the corresponding partition line segments are the crosscuts Γ_1 and Γ_4. Since the polynomial p_0 contributes $\binom{d+2}{2}$ free parameters, the dimension of $S_d^r(\Delta, \Gamma)$ is now shown to be

$$\binom{d+2}{2} + 2\binom{d-r+1}{2} + C_d^r(3) + C_d^r(4)$$

which is the formula (4.6) with $L = V = 2$, $N_1 = 3$, and $N_2 = 4$.

The procedure discussed in the above example can be used to establish Theorem 4.3, assuming of course that (4.7) is valid (cf. [73]).

It is important to observe that if Δ is a purely crosscut partition, and by this we mean that all partition line segments are crosscuts of Ω, then a basis of the bivariate spline space $S_d^r(\Delta, \Omega)$ is given by the union of the following three collections of functions: (i) a basis of the polynomial space π_d^2, (ii) the set of "crosscut functions" defined to be zero on one component of the complement of Γ_i relative to Ω and ℓ_i^{r+1} on the other component, as shown in Figure 4.4, where the Γ_i's are the crosscuts of the partition of the simply connected region Ω, and (iii) the set of all "bivariate truncated powers," defined by using the smoothing cofactors $\tilde{q}_1, \hat{q}_1, \cdots, \tilde{q}_{N_i}, \hat{q}_{N_i}$, at each interior vertex A_i, where $q_j = \hat{q}_j + \tilde{q}_j$, $j = 1, \cdots, N_i$, are determined by the conformality condition

$$\sum_{j=1}^{N_i} q_j \ell_{i,j}^{r+1} = 0, \quad q_1, \cdots, q_{N_i} \in \pi_{d-r-1}^2,$$

with supports properly contained in a half-plane as shown in Figure 4.5 below. This is accomplished by setting the smoothing cofactors $\hat{q}_1, \cdots, \hat{q}_{N_i}$ to be zero, so that $q_j = \tilde{q}_j + \hat{q}_j = \tilde{q}_j$ for all $j = 1, \cdots, N_i$. For more details, the reader is referred to Chui and Wang [73].

We remark that the terminology of "bivariate truncated powers" is used here to remind ourselves of the multivariate truncated powers discussed in §2.3. The totality of all crosscut functions and bivariate truncated powers is called the collection of generalized truncated powers (and for the special case where $N_i = 3$ for all i, *cone splines*, in de Boor and Höllig [31]).

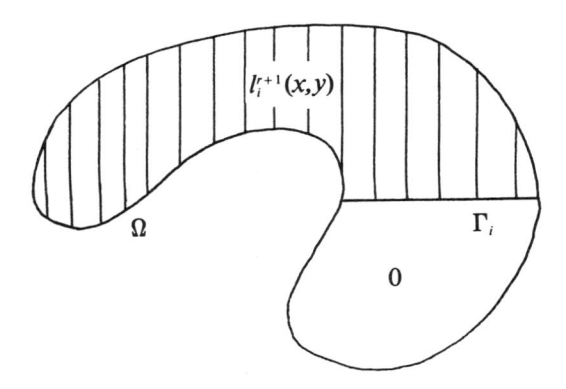

FIG. 4.4

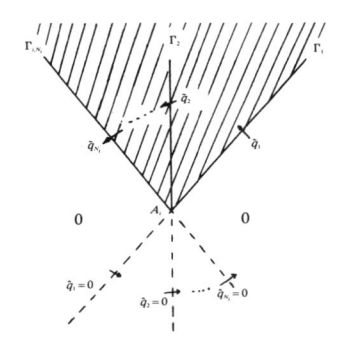

FIG. 4.5

In the study of spline functions in more than one variable, one expects many interesting properties that cannot be found in univariate splines. One such property is that a C^r piecewise polynomial function may have higher order derivatives on the lower dimensional grid partition. For example, if $s = 3$, then C^r splines may be C^{2r} on edges and C^{4r} at vertices. To be more precise, let Δ be a grid partition of an s-dimensional region Ω. The partition consists of j-dimensional algebraic manifolds, $j = 0, \cdots, s-1$, where a grid point is considered to be a zero-dimensional manifold. The space $\widehat{S}_d^r(\Delta) = \widehat{S}_d^r(\Delta, \Omega)$ of functions $f \in C^r(\Omega)$, which have $(2^{s-j-1}r)$th-order derivatives at the j-dimensional manifold of Δ, $j = 0, \cdots, s-1$, such that the restrictions of f on each subregion are in π_d^s, will be called the super spline subspace of $S_d^r(\Delta, \Omega)$. This terminology is introduced in Chui and Lai [66] in constructing multivariate vertex splines, a topic to be discussed in Chapter 6. It turns out that in approximation from the spline space $S_d^r(\Delta, \Omega)$, $\Omega \subset \mathbf{R}^s$, when the degree d is high enough, say $d \geq 2^s r + 1$, the subspace $\widehat{S}_d^r(\Delta, \Omega)$ of super splines gives the same order of approximation as the spline space $S_d^r(\Delta, \Omega)$.

We now return to restricting our attention to bivariate splines. A bivariate super spline in $S_d^r(\Delta)$ has derivatives up to at least order $2r$ at each interior grid point. Again let Ω be a simply connected region in $\mathbf{R}^2$ and partition Ω by drawing in straight line segments which are either crosscuts or rays. The following dimension result is obtained in Chui and He [59].

THEOREM 4.4. *Let Δ be a quasi-crosscut partition of a simply connected region Ω in $\mathbf{R}^2$ with L crosscuts and V interior vertices $A_1, \cdots, A_V$, such that for each $i = 1, \cdots, V$, the number of crosscuts or rays passing through or starting at A_i is a finite number N_i. Then the dimension of the bivariate super spline space $\widehat{S}_d^r(\Delta)$ is*

$$(4.8) \qquad \dim \widehat{S}_d^r(\Delta) = \binom{d+2}{2} + L E_d^r + \sum_{i=1}^{V} \widehat{C}_d^r(N_i)$$

where

$$(4.9) \qquad E_d^r = \binom{d-r+1}{2} - 2\binom{r+1}{2} + \binom{(3r-d+1)_+}{2}$$

and

$$(4.10) \qquad \widehat{C}_d^r(n) = \sum_{j=r+1}^{d-r} (r+j+1-n(d-2r))_+$$
$$+ n\left(\binom{d-r+1}{2} - 2\binom{r+1}{2} + \binom{(3r-d+1)_+}{2}\right)$$
$$- \binom{d+2}{2} + \binom{2r+2}{2}.$$

Note that for $d \geq 3r$, we have

$$E_d^r = \binom{d-r+1}{2} - 2\binom{r+1}{2}$$

and for either $d \geq 3r+1$ and $n \geq 3$, or $d \geq 4r+1$, we have

$$\widehat{C}_d^r(n) = n\left[\binom{d-r+1}{2} - 2\binom{r+1}{2}\right] - \binom{d+2}{2} + \binom{2r+2}{2}.$$

4.3. Upper and lower bounds on dimensions. If the partition Δ is not a quasi-crosscut partition, then the dimensions of the bivariate spline space and its super spline subspace are usually very difficult to determine. We will only consider the important case where the partition is in fact a triangulation.

Let Ω be a closed simply connected polygonal region in $\mathbf{R}^2$ and $\Delta = \{\Omega_j\}$ be a finite collection of closed triangular regions such that

(i) $\Omega = \cup \, \Omega_j$,

(ii) for $i \neq j$, $\Omega_i \not\subseteq \Omega_j$ and $\Omega_j \not\subseteq \Omega_i$,

and

(iii) for $i \neq j$, either $\Omega_i \cap \Omega_j = \emptyset$, or Ω_i, Ω_j share a common vertex or a common edge.

We will say that Δ is a (regular) triangulation of Ω. As before, let

$$S_d^r(\Delta) = \{f \in C^r(\Omega): \ f|_{\Omega_i} \in \pi_d^2, \text{ for each } i\}$$

and

$$\widehat{S}_d^r(\Delta) = \{f \in S_d^r(\Delta): f^{(i)}(A) \text{ exists for } i = 0, \cdots, 2r$$
$$\text{and any vertex } A \text{ of } \Delta\}.$$

We are interested in studying the dimension of $S_d^r(\Delta)$ and its super spline subspace $\widehat{S}_d^r(D)$.

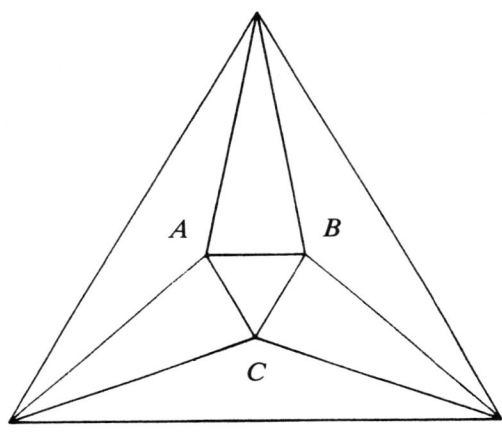

FIG. 4.6

We first remark that it is in general not possible to give the exact dimension formula, since such a formula might have to depend on the geometric configurations of the partition. The first example is given in Morgan and Scott [160], where it is shown that for the triangulation Δ_s in Figure 4.6, the dimension of $S_2^1(\Delta_s)$ is either 7 or 6 depending on whether the triangle ABC is perfectly symmetric or not. Several other interesting examples are given in Gmelig Meyling and Pfluger [119]. For this reason, we must be satisfied with good upper and lower bounds of the dimension values.

Let E and V denote the number of *interior* edges and vertices of the triangulation Δ, respectively. For instance, we may label the vertices as $A_1, \cdots, A_V$. For each $i = 1, \cdots, V$, let N_i denote the number of edges with different slopes attached to A_i, and $\widetilde{N}_i$ the number of edges with different slopes attached to A_i but not A_j, where $j < i$. Set

$$(4.11) \qquad \sigma_i = \sum_{j=1}^{d-r}(r+j+1-jN_i)_+$$

and

$$(4.12) \qquad \tilde{\sigma}_i = \sum_{j=1}^{d-r}(r+j+1-j\widetilde{N}_i)_+.$$

Note that $\tilde{\sigma}_i \leq \sigma_i$, and $\tilde{\sigma}_i$ depends on the particular ordering of the vertices. In Schumaker [185] and [188], respectively, both a lower bound and an upper bound for $\dim S_d^r(\Delta)$ are obtained. These results can be summarized as follows:

THEOREM 4.5. *Let*

$$(4.13) \qquad \begin{aligned} D = \dim S_d^r(\Delta) &- \binom{d+2}{2} - \binom{d-r+1}{2}E \\ &+ \left[\binom{d+2}{2} - \binom{r+2}{2}\right]V. \end{aligned}$$

Then

$$(4.14) \qquad \sum_{i=1}^{V}\sigma_i \leq D \leq \sum_{i=1}^{V}\tilde{\sigma}_i.$$

To arrive at the lower bound of $\dim S_d^r(\Delta)$, we first verify the special case where $V = 1$. This is then a quasi-crosscut partition, and it can be shown that the lower bound value is actually attained. The general result on lower bound can then be verified by mathematical induction, using the so-called "merging" procedure. The upper bound result is obtained by constructing a set of linear functionals with cardinality given by the upper bound value so that the zero function is the only spline in $S_d^r(\Delta)$ that lies in the kernel.

From the nature of the proof, it seems that the lower bound has a better chance of giving the exact dimension. In fact, in Hong [135] it is proved that for $d \geq 3r + 2$, the dimension of $S_d^r(\Delta)$ is given by the lower bound value; that is,

$$\dim S_d^r(\Delta) = \binom{d+2}{2} + \binom{d-r+1}{2} E$$
$$- \left[\binom{d+2}{2} - \binom{r+2}{2}\right] V + \sum_{i=1}^{V} \sigma_i.$$

To discuss the analogous results for super spline subspaces, let

(4.15) $$\sigma_i^s = \sum_{j=r+1}^{d-r} (r + j + 1 - (d - 2r)N_i)_+$$

and

(4.16) $$\tilde{\sigma}_i^s = \sum_{j=r+1}^{d-r} (r + j + 1 - (d - 2r)\tilde{N}_i)_+ .$$

The following result is obtained in Chui and He [59].

THEOREM 4.6. *Let*

(4.17) $$D^s = \dim \widehat{S}_d^r(\Delta) - \binom{d+2}{2}$$
$$- \left[\binom{d-r+1}{2} - 2\binom{r+1}{2} + \binom{(3r+1-d)_+}{2}\right] E$$
$$+ \left[\binom{d+2}{2} - \binom{2r+2}{2}\right] V.$$

Then

(4.18) $$\sum_{i=1}^{V} \sigma_i^s \leq D^s \leq \sum_{i=1}^{V} \tilde{\sigma}_i^s .$$

Again, we believe that the lower bound has a better chance of giving the actual dimension.

4.4. Approximation order. In §3.4, we discussed some interesting results on the order of approximation from the bivariate spline spaces on the three- and four-directional meshes. We now turn to the study of an arbitrary (regular) triangulation Δ, satisfying (i)-(iii) in §4.3. Let

$$h = \max_j (\text{diam } \Omega_j)$$

where the Ω_j's are the triangular regions that define the triangulation Δ. Of course, the approximation order m is defined in the form of $O(h^m)$. It is clear that not much can be said if the spline space $S_d^r(\Delta)$ is required to be "too smooth" (or d is not much larger than r). On the other hand, if $d \geq 4r+1$, it is well known (cf. Ženíček [205], [206]) that the full order $m = d+1$ of approximation is attained. In fact, even the super spline subspace $\widehat{S}_d^r(\Delta)$ provides this approximation order and vertex splines (to be discussed in Chapter 6), which have smallest supports, can be constructed to facilitate the approximation and interpolation schemes. de Boor and Höllig [32] recently proved that, even for $d \geq 3r+2$, the full approximation order can be obtained.

THEOREM 4.7. *Let $d \geq 3r+2$. Then there exists a positive constant C which depends only on d and the smallest angle in the triangulation Δ, such that*

$$\text{dist}_{L^\infty}(f, S_d^r(\Delta)) \leq C \parallel D^{d+1} f \parallel_\Omega h^{d+1}$$

for all $f \in C^{d+1}(\Omega)$.

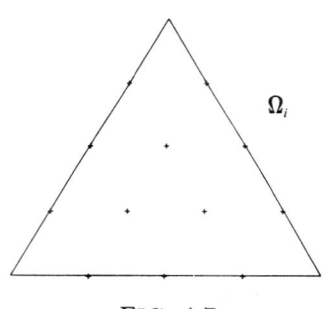

FIG. 4.7

In proving this theorem, the set Λ_i of the three vertices of each triangle Ω_i is "refined" by adding $\binom{d}{2} - 3$ equally spaced extra points on Ω_i as in Figure 4.7, yielding a set $\overline{\Lambda}_i$ of $\binom{d}{2}$ equally spaced points on Ω_i.

Hence, for each $f \in C(\Omega)$, there exists a unique spline $Pf \in S_d^0(\Delta)$ defined by $(Pf - f)(\mathbf{x}) = 0$ for all $\mathbf{x} \in \overline{\Lambda}_i, i = 1, \cdots, V$. Since it is clear that for $f \in C^{d+1}(\Omega)$,

$$\parallel f - Pf \parallel_\Omega \leq C_1(1+ \parallel P \parallel) \parallel D^{d+1} f \parallel_\Omega h^{d+1}$$

where C_1 is some positive constant depending only on $\triangle$, we have

$$\text{(4.19)} \quad \text{dist}_{L^\infty}(f, S_d^r(\triangle)) \leq C_1(1+ \|P\|) \|D^{d+1}f\|_\Omega h^{d+1} + \text{dist}_{L^\infty}(Pf, S_d^r(\triangle)).$$

That is, the approximation order of $S_d^r(\triangle)$ strictly depends on how well Pf, where f is sufficiently smooth, can be approximated from $S_d^r(\triangle)$. For this purpose, the Hahn–Banach theorem gives the formula

$$\text{(4.20)} \quad \begin{aligned}&\text{dist}_{L^\infty}(Pf, S_d^r(\triangle))\\ &= \max\{|\lambda Pf|: \ \lambda \in (S_d^r(\triangle))^\perp, \|\lambda\|=1\}\end{aligned}$$

with

$$\text{(4.21)} \quad (S_d^r(\triangle))^\perp = \{\lambda \in (S_d^0(\triangle))^*: \ \lambda g = 0, \text{ all } g \in S_d^r(\triangle)\}.$$

To complete the proof, de Boor and Höllig [32] supply a "good" basis for $(S_d^r(\triangle))^\perp$ and the argument is fairly complicated.

Recently, Chui and Lai [CAT Report # 164, 1988] identified a "superspline" subspace of $S_d^r(\triangle), d \geq 3r+2$, and gave a quasi-interpolation formula, which is based on a basis consisting of vertex splines of this subspace, that also gives the full approximation order $d+1$. If $d \geq 4r+1$, much simpler quasi-interpolation formulas are given in Chapter 6.

We also remark that value $3r+2$ is sharp in the sense that if $d = 3r+1$, the full order of approximation is at most d, even for the three-directional mesh. This is shown for $r = 1, 2, 3$ in [32] and in general by Jia [Private Communication]. To see this, we may appeal to (4.20) and exhibit a smooth function f (a monomial, say) and a $\lambda \in (S_{3r+1}^r(\triangle))^\perp$ such that

$$\frac{|\lambda Pf|}{\|\lambda\|} \geq C_f h^{3r+1}$$

for some positive constant C_f dependent only on f.

4.5. Other subspaces. We first briefly touch on the topic of bivariate spline subspaces with preassigned boundary conditions. In addition to its important applications, information on dimension and basis structure helps the study of minimally and quasi-minimally supported splines. So far only the zero boundary condition has been considered, and the only known results are concerned with very specific situations. For instance, Chui and Schumaker [69] give the dimension and construct a local basis for the rectangular partition; and Chui, Schumaker, and Wang [71] and [72] obtain analogous results for $S_3^1(\triangle_{MN}^{(1)})$ and $S_2^1(\triangle_{MN}^{(2)})$, respectively. One interesting result for $S_3^1(\triangle_{MN}^{(1)})$ is worth mentioning. Although we have assigned the zero boundary condition, it is shown in [71] that any basis of this subspace has to contain a global element; that is, one whose support extends from the left to the right boundaries or from the bottom to the top boundaries of Ω_{MN}. The results on $S_2^1(\triangle_{MN}^{(2)})$ are not surprising, and in fact, have straightforward generalization to nonuniform type-2 triangulation given by Wang and He [203].

Another interesting subspace is the space of periodic bivariate splines. In ter Morsche [161] the periodic spline subspace of $S_3^1(\triangle_{MN}^{(1)})$ is studied and the property of unisolvence in interpolating periodic data is also discussed. The periodicity requirement leads naturally to approximation of Fourier coefficients of a periodic function by interpolating bivariate periodic splines. This topic is studied in ter Morsche [161], and we will delay our brief discussion of it to Chapter 9.

CHAPTER 5

Bézier Representation and Smoothing Techniques

In this chapter we will consider only grid partitions consisting of simplices and parallelepipeds. For these geometric subregions it is more convenient to use barycentric coordinates than the usual Cartesian coordinates, since each polynomial piece of a multivariate spline may then take on the Bézier representation.

5.1. Bézier polynomials. Let $\mathbf{x}^0, \cdots, \mathbf{x}^s \in \mathbf{R}^s, s \geq 1$. The simplex

$$\langle \mathbf{x}^0, \cdots, \mathbf{x}^s \rangle = \left\{ \sum_{i=0}^s \lambda_i \mathbf{x}^i : \sum_{i=0}^s \lambda_i = 1, \lambda_i \geq 0 \right\},$$

with vertices $\mathbf{x}^0, \cdots, \mathbf{x}^s$ is called an s-simplex, if its (signed) volume

$$\text{vol}_s \langle \mathbf{x}^0, \cdots, \mathbf{x}^s \rangle = \frac{1}{s!} \begin{vmatrix} 1 & x_1^0 & \cdots & x_s^0 \\ \vdots & \vdots & & \vdots \\ 1 & x_1^s & \cdots & x_s^s \end{vmatrix}$$

is nonzero. Here, $\mathbf{x}^i = (x_1^i, \cdots, x_s^i)$. Suppose that $\langle \mathbf{x}^0, \cdots, \mathbf{x}^s \rangle$ is an s-simplex. Then any $\mathbf{x} = (x_1, \cdots, x_s)$ in $\mathbf{R}^s$ can be identified by an $(s+1)$-tuple $(\lambda_0, \cdots, \lambda_s)$, where

$$\lambda_i = \lambda_i(\mathbf{x}) = \frac{\text{vol}_s \langle \mathbf{x}^0, \cdots, \mathbf{x}^{i-1}, \mathbf{x}, \mathbf{x}^{i+1}, \cdots, \mathbf{x}^s \rangle}{\text{vol}_s \langle \mathbf{x}^0, \cdots, \mathbf{x}^s \rangle}.$$

This $(s+1)$-tuple is called the barycentric coordinate of $\mathbf{x}$ relative to the s-simplex $\langle \mathbf{x}^0, \cdots, \mathbf{x}^s \rangle$. Note that each $\lambda_i = \lambda_i(\mathbf{x})$ is a linear polynomial in $\mathbf{x}$. For any $\beta = (\beta_0, \cdots, \beta_s) \in \mathbf{Z}_+^{s+1}$, and $n \in \mathbf{Z}_+$, we will use the usual multivariate notation

$$\lambda^\beta = \lambda_0^{\beta_0} \cdots \lambda_s^{\beta_s}, \quad \beta! = \beta_0! \cdots \beta_s!, \quad |\beta| = \beta_0 + \cdots + \beta_s.$$

Hence,

(5.1) $$\phi_\beta^n(\lambda) = \frac{n!}{\beta!} \lambda^\beta, \quad |\beta| = n,$$

is a polynomial in π_n^s. In fact, it is easy to see that $\{\phi_\beta^n(\lambda): |\beta| = n\}$ is a basis of the polynomial space π_n^s. The representation

(5.2) $$P_n(\mathbf{x}) = \sum_{|\beta|=n} a_\beta^n \phi_\beta^n(\lambda)$$

is called a *Bézier polynomial* of total degree n relative to the s-simplex $\langle \mathbf{x}^0, \cdots, \mathbf{x}^s \rangle$. In addition, the set

$$\{(\frac{\beta_0}{n}\mathbf{x}^0 + \cdots + \frac{\beta_s}{n}\mathbf{x}^s, a_\beta^n): |\beta| = n\},$$

which we will write $\{a_\beta^n\} = \{a_\beta^n: |\beta| = n\}$ for brevity, is called the *Bézier net* of the polynomial P_n. Hence, to describe the polynomial P_n, we simply write down its Bézier net, and in the case $s = 2$, we may simply show this Bézier net on a triangular array as in Figure 5.1.

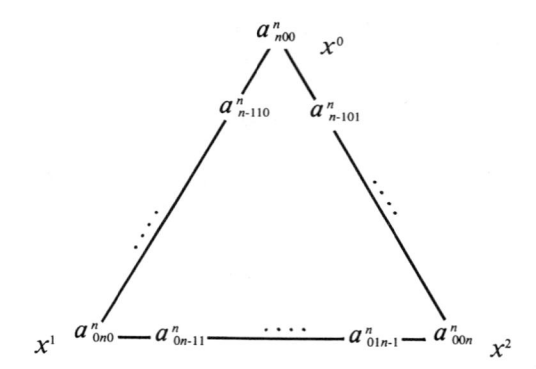

FIG. 5.1

Next, let $\mathbf{w}^1, \cdots, \mathbf{w}^{2^s} \in \mathbf{R}^s$ be vertices of an s-parallelepiped given by the convex hull

$$\langle \mathbf{w}^1, \cdots, \mathbf{w}^{2^s} \rangle$$

having positive s-dimensional volume and $2s$ boundary faces $A_1, \cdots, A_{2s}$ with positive $(s-1)$-dimensional Lebesgue measure, so that A_{2k-1} is parallel to A_{2k} for $k = 1, \cdots, s$, say. In addition, the vertices are labeled in such a manner that the edges of the s-parallelepiped are parallel to the vectors $\mathbf{w}^{i+1} - \mathbf{w}^1$, $i = 1, \cdots, s$, so that

$$\langle \mathbf{w}^1, \cdots, \mathbf{w}^{2^s} \rangle = \{\sum_{i=1}^s t_i(\mathbf{w}^{i+1} - \mathbf{w}^1): 0 \leq t_1, \cdots, t_s \leq 1\}.$$

BÉZIER REPRESENTATION

For each $\mathbf{x} \in \langle \mathbf{w}^1, \cdots, \mathbf{w}^{2^s} \rangle$, and $k = 1, \cdots, s$, it is clear that

$$\text{vol}_s \langle A_{2k-1}, \mathbf{x} \rangle + \text{vol}_s \langle A_{2k}, \mathbf{x} \rangle = \frac{1}{s} \text{vol}_s \langle \mathbf{w}^1, \cdots, \mathbf{w}^{2^s} \rangle$$

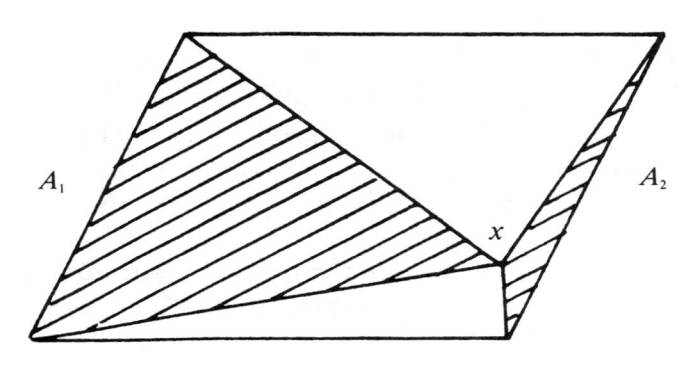

FIG. 5.2

(cf. Figure 5.2). Hence, the linear polynomials

(5.3) $$\mu_k = \mu_k(\mathbf{x}) = s \frac{\text{vol}_s \langle A_{2k-1}, \mathbf{x} \rangle}{\text{vol}_s \langle \mathbf{w}^1, \cdots, \mathbf{w}^{2^s} \rangle}$$

satisfy: $0 \leq \mu_k \leq 1$, $\mu_k(\mathbf{x}) = 0$ if and only if $\mathbf{x} \in A_{2k-1}$, and $\mu_k(\mathbf{x}) = 1$ if and only if $\mathbf{x} \in A_{2k}$. Set $\mu = (\mu_1, \cdots, \mu_s)$. Then we may consider polynomials $\widetilde{P}_\alpha(\mathbf{x})$ of coordinate degree $\alpha = (\alpha_1, \cdots, \alpha_s) \in \mathbf{Z}_+^s$ in the form:

(5.4) $$\widetilde{P}_\alpha(\mathbf{x}) = \sum_{\gamma \leq \alpha} \widetilde{a}_\gamma^\alpha \widetilde{\phi}_\gamma^\alpha(\mu)$$

where $\gamma = (\gamma_1, \cdots, \gamma_s)$ and

(5.5) $$\widetilde{\phi}_\gamma^\alpha(\mu) = \binom{\alpha}{\gamma} \mu^\gamma (1-\mu)^{\alpha-\gamma}$$

with

$$\binom{\alpha}{\gamma} = \binom{\alpha_1}{\gamma_1} \cdots \binom{\alpha_s}{\gamma_s}$$

and

$$(1-\mu)^{\alpha-\gamma} = (1-\mu_1)^{\alpha_1-\gamma_1} \cdots (1-\mu_s)^{\alpha_s-\gamma_s}.$$

If $\alpha = (n, \cdots, n), n \in Z_+$, we will simply write

$$\widetilde{P}_n = \widetilde{P}_\alpha, \; \widetilde{\phi}_\gamma^n(\mu) = \widetilde{\phi}_\gamma^{(n,\cdots,n)}(\mu), \; \widetilde{a}_j^n = \widetilde{a}_\gamma^{(n,\cdots,n)}.$$

The polynomial $\widetilde{P}_\alpha(x)$ in (5.4) is called a Bézier polynomial of coordinate degree α relative to the parallelepiped $\langle \mathbf{w}^1, \cdots, \mathbf{w}^{2^s} \rangle$, and the set

$$\left\{ \left(\mathbf{w}^1 + \frac{\gamma_1}{\alpha_1}(\mathbf{w}^2 - \mathbf{w}^1) + \cdots + \frac{\gamma_s}{\alpha_s}(\mathbf{w}^{s+1} - \mathbf{w}^1), \tilde{a}_\gamma^\alpha \right) \right\}$$

which will be written as $\{\tilde{a}_\gamma^\alpha\}$ for brevity, is called the Bézier net of $\widetilde{P}_\alpha$ relative to this parallelepiped. For more details, see Chui and Lai [66].

Note that for $s = 1$, both the simplex and parallelepiped settings reduce to an interval and the polynomials P_n and $\widetilde{P}_\alpha$ become a Bernstein polynomial on this interval.

5.2. Smoothness conditions for adjacent simplices. We will first discuss the smoothness conditions for two adjacent simplices (see Figure 5.3).

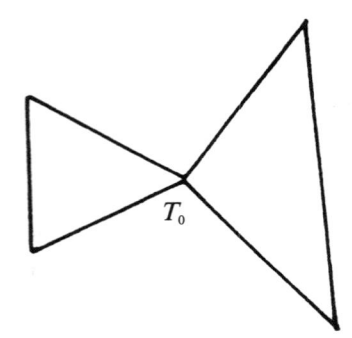

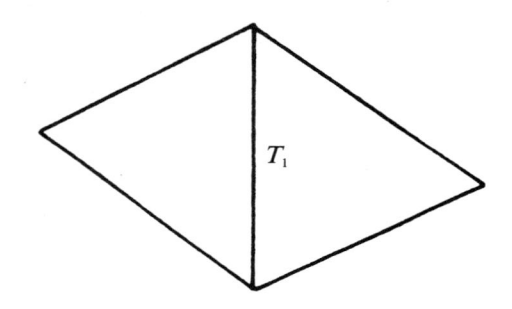

FIG. 5.3

Let $T_k = \langle \mathbf{x}^0, \cdots, \mathbf{x}^k \rangle$ be a k-simplex in $\mathbf{R}^s$ where $0 \le k < s$, and let

$$S_a = \langle \mathbf{x}^0, \cdots, \mathbf{x}^k, \mathbf{x}^{k+1}, \cdots, \mathbf{x}^s \rangle$$
$$S_b = \langle \mathbf{x}^0, \cdots, \mathbf{x}^k, \mathbf{y}^{k+1}, \cdots, \mathbf{y}^s \rangle$$

be two adjacent s-simplices with $S_a \cap S_b = T_k$. Suppose that we have a function $F(\mathbf{x})$ in $C(S_a \cup S_b)$ with restrictions:

$$F|_{S_a}(\mathbf{x}) = P_n(\mathbf{x}) = \sum_{|\beta|=n} a_\beta^n \phi_\beta^n(\lambda)$$
$$F|_{S_b}(\mathbf{x}) = Q_n(\mathbf{x}) = \sum_{|\beta|=n} b_\beta^n \phi_\beta^n(\eta)$$

where $\lambda = (\lambda_0, \cdots, \lambda_s)$ and $\eta = (\eta_0, \cdots, \eta_s)$ are the barycentric coordinates of $\mathbf{x}$ relative to S_a and S_b, respectively. Note that $\mathbf{x} \in T_k$ if and

only if
$$\mathbf{x} = \sum_{i=0}^{k} \lambda_i \mathbf{x}^i = \sum_{i=0}^{k} \eta_i \mathbf{x}^i,$$
so that $\lambda_i = \eta_i, i = 0, \cdots, k$ on T_k.

As in §2.2, we will use the notation
$$D_\mathbf{y} = \sum_{i=1}^{s} y_i \frac{\partial}{\partial x_i}$$
where $\mathbf{x} = (x_1, \cdots, x_s)$ and $\mathbf{y} = (y_1, \cdots, y_s)$; but to simplify this notation when $\mathbf{y} = \mathbf{x}^i - \mathbf{x}^j$, say, we write
$$D_{ij} = D_{\mathbf{x}^i - \mathbf{x}^j}, \quad i \neq j.$$

For differences, we use the notation
$$s_i \alpha = (\alpha_0, \cdots, \alpha_{i-1}, \alpha_i + 1, \alpha_{i+1}, \cdots, \alpha_s)$$
where $\alpha = (\alpha_0, \cdots, \alpha_s)$ and
$$\triangle_{ij} a_\alpha^n = a_{s_i \alpha}^n - a_{s_j \alpha}^n.$$

We have the following relationship between D_{ij} and $\triangle_{ij}$ for Bézier polynomials on simplices (cf. Chui and Lai [64]).

LEMMA 5.1. For $i \neq j$,
$$(5.6) \qquad (D_{ij} P_n)(\mathbf{x}) = n \sum_{|\alpha| = n-1} \triangle_{ij} a_\alpha^n \phi_\alpha^{n-1}(\lambda).$$

To prove this lemma, we write $\mathbf{x}^i = (x_1^i, \cdots, x_s^i)$ and $\mathbf{x} = (x_1, \cdots, x_s)$. Since
$$x_\ell = \sum_{t=0}^{s} \lambda_t x_\ell^t, \quad \ell = 1, \cdots, s,$$
we have
$$(D_{ij} P_n)(\mathbf{x}) = \sum_{\ell=1}^{s} (x_\ell^i - x_\ell^j) \frac{\partial}{\partial x_\ell} P_n(\mathbf{x})$$
$$= \left(\frac{\partial}{\partial \lambda_i} - \frac{\partial}{\partial \lambda_j} \right) P_n(\mathbf{x}),$$

and (5.6) follows from a simple change of indices in

$$\sum_{\alpha_0+\cdots+\alpha_s=n} \frac{n!}{\alpha_0!\cdots\alpha_s!} a_\alpha^n \left(\frac{\partial}{\partial \lambda_i} - \frac{\partial}{\partial \lambda_j}\right)(\lambda_0^{\alpha_0}\cdots\lambda_s^{\alpha_s}).$$

In the following, we will use the notation

(5.7) $$c_{ji} = \frac{\text{vol}_s\langle \mathbf{x}^0,\cdots,\mathbf{x}^{i-1},\mathbf{y}^j,\mathbf{x}^{i+1},\cdots,\mathbf{x}^s\rangle}{\text{vol}_s\langle \mathbf{x}^0,\cdots,\mathbf{x}^s\rangle}.$$

This means, of course, that $\mathbf{y}^j = \sum_{i=0}^s c_{ji}\mathbf{x}^i$. We have the following necessary and sufficient conditions that the Bézier nets of $P_n(\mathbf{x})$ and $Q_n(\mathbf{x})$ must satisfy so that $F \in C^r(S_a \cup S_b)$.

THEOREM 5.1. *Let $r \in \mathbf{Z}_+$. Then $F \in C^r(S_a \cup S_b)$ if and only if*

(5.8) $$\Delta_{k+1,0}^{\gamma_{k+1}}\cdots\Delta_{s0}^{\gamma_s} b^n_{(\alpha_0,\cdots,\alpha_k,0,\cdots,0)}$$

$$= \left(\sum_{i=1}^s c_{k+1,i}\Delta_{i0}\right)^{\gamma_{k+1}}\cdots\left(\sum_{i=1}^s c_{si}\Delta_{i0}\right)^{\gamma_s} a^n_{(\alpha_0,\cdots,\alpha_k,0,\cdots,0)}$$

for all $\gamma_{k+1}+\cdots+\gamma_s = \ell$, $\alpha_0+\cdots+\alpha_k = n-\ell$, *and* $\ell = 0,\cdots,r$. *In particular*, $F \in C(S_a \cup S_b)$ *if and only if* $b^n_{(\alpha_0,\cdots,\alpha_k,0,\cdots,0)} = a^n_{(\alpha_0,\cdots,\alpha_k,0,\cdots,0)}$ *for all* $\alpha_0 + \cdots + \alpha_k = n$.

The formulation of (5.8) is given in Chui and Lai [66]. Other formulations can be found in de Boor [22] and Farin [105], [108]. The proof of Theorem 5.1 is a consequence of setting

$$D_{k+1,0}^{\gamma_{k+1}}\cdots D_{s0}^{\gamma_s} P_n(\mathbf{x}) = D_{k+1,0}^{\gamma_{k+1}}\cdots D_{s0}^{\gamma_s} Q_n(\mathbf{x})$$

for $\mathbf{x} \in T_k$, where $\gamma_{k+1} + \cdots + \gamma_s = \ell$ and $\ell = 0,\cdots,r$, and applying Lemma 5.1, using the fact that

$$\mathbf{y}^j = \sum_{i=0}^s c_{ji}\mathbf{x}^i.$$

Indeed, since

$$\mathbf{y}^j - \mathbf{x}^0 = \sum_{i=1}^s c_{ji}(\mathbf{x}^i - \mathbf{x}^0),$$

we have, by Lemma 5.1,

$$(D^\gamma_{\mathbf{y}^j-\mathbf{x}^0})P_n(\mathbf{x}) = \left(\sum_{i=1}^s c_{ji}D_{i0}\right)^\gamma P_n(\mathbf{x})$$

$$= \frac{n!}{(n-\gamma)!} \sum_{|\alpha|=n-\gamma} \left(\sum_{i=1}^s c_{ji}\Delta_{i0}\right)^\gamma a^n_\alpha \phi^{n-\gamma}_\alpha(\lambda)$$

and

$$(D^\gamma_{\mathbf{y}^j-\mathbf{x}^0})Q_n(\mathbf{x}) = \frac{n!}{(n-\gamma)!} \sum_{|\alpha|=n-\gamma} \Delta^\gamma_{j0} b^n_\alpha \phi^{n-\gamma}_\alpha(\eta).$$

Hence, by taking all mixed directional derivatives of total order r and equating the two expressions with one another on the common boundary T_k, we have the necessary and sufficient condition (5.8) for $F(\mathbf{x})$ to be in $C^r(S_a \cup S_b)$.

Example 5.1. Consider $s = 2, n = 3$, and $k = 0$ (cf. Figure 5.4).

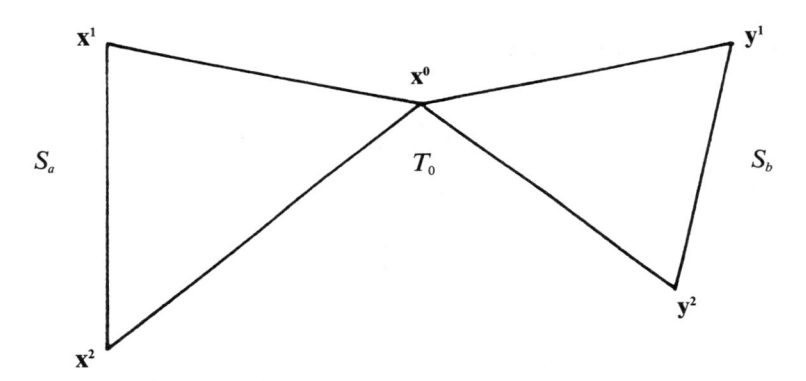

FIG. 5.4

We have:
(i) $F \in C^0(S_a \cup S_b)$ if and only if

$$(a) \quad b_{300} = a_{300}$$

is satisfied.

(ii) $F \in C^1(S_a \cup S_b)$ if and only if both (a) and

$$(b) \quad \begin{cases} b_{210} = c_{10}a_{300} + c_{11}a_{210} + c_{12}a_{201} \\ b_{201} = c_{20}a_{300} + c_{21}a_{210} + c_{22}a_{201} \end{cases}$$

are satisfied.

(iii) $F \in C^2(S_a \cup S_b)$ if and only if (a), (b), and

(c) $\begin{cases} b_{120} = c_{10}b_{210} + c_{11}(c_{10}a_{210} + c_{11}a_{120} + c_{12}a_{111}) \\ \qquad\quad + c_{12}(c_{10}a_{201} + c_{11}a_{111} + c_{12}a_{102}) \\ b_{111} = c_{20}b_{210} + c_{21}(c_{10}a_{210} + c_{11}a_{120} + c_{12}a_{111}) \\ \qquad\quad + c_{22}(c_{10}a_{201} + c_{11}a_{111} + c_{12}a_{102}) \\ b_{102} = c_{20}b_{201} + c_{21}(c_{20}a_{210} + c_{21}a_{121} + c_{22}a_{111}) \\ \qquad\quad + c_{22}(c_{20}a_{201} + c_{21}a_{111} + c_{22}a_{102}) \end{cases}$

are satisfied.

Of course, for $n = 3$ (that is, cubic polynomials), $F \in C^3(S_a \cup S_b)$ if and only if $P_3(\mathbf{x}) \equiv Q_3(\mathbf{x})$.

The geometry in $\mathbf{R}^3$ of the smoothness conditions is quite nice. For instance, $F \in C^1(S_a \cup S_b)$ if and only if the two triangles in $\mathbf{R}^3$ formed by the Bézier nets $\{a_{300}, a_{210}, a_{120}\}$ and $\{b_{300}, b_{210}, b_{201}\}$, are coplanar. See Figure 5.5.

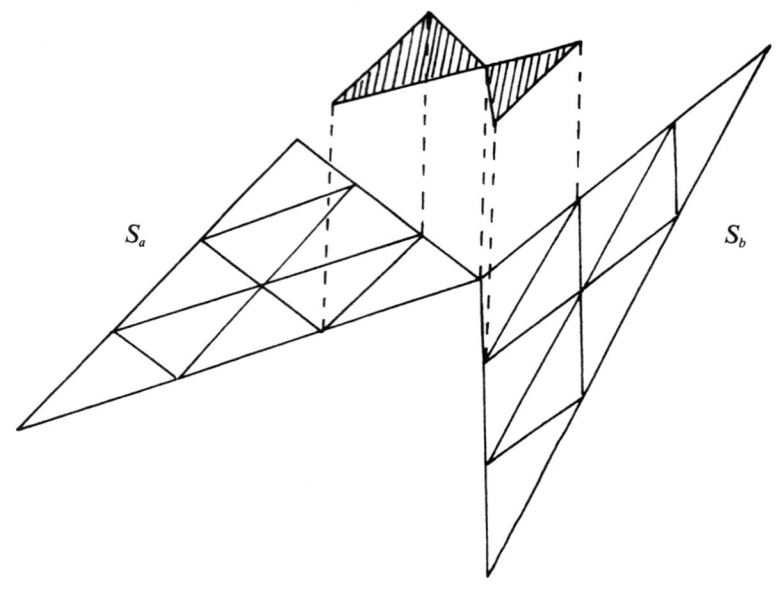

FIG. 5.5

Example 5.2. Consider $s = 2, n = 3$, and $k = 1$, with $T_1 = \langle \mathbf{x}^0, \mathbf{x}^1 \rangle$ (cf. Figure 5.6).

BÉZIER REPRESENTATION

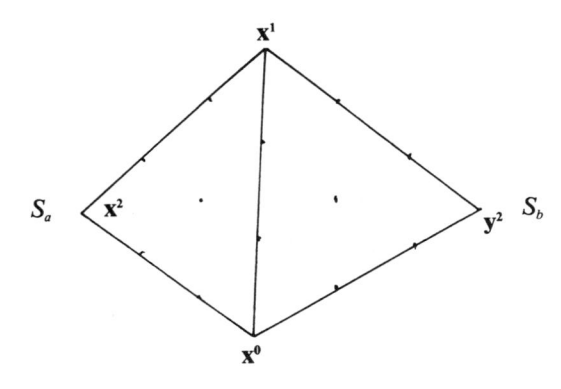

FIG. 5.6

We have:
(i) $F \in C^0(S_a \cup S_b)$ if and only if

(a) $b_{300} = a_{300}, \quad b_{210} = a_{210}, \quad b_{120} = a_{120}, \quad b_{030} = a_{030}$

is satisfied.
(ii) $F \in C^1(S_b \cup S_b)$ if and only if both (a) and

(b) $\begin{cases} b_{201} = c_{20}a_{300} + c_{21}a_{210} + c_{22}a_{201} \\ b_{111} = c_{20}a_{210} + c_{21}a_{120} + c_{22}a_{111} \\ b_{021} = c_{20}a_{120} + c_{21}a_{030} + c_{22}a_{021} \end{cases}$

are satisfied.
(iii) $F \in C^2(S_a \cup S_b)$ if and only if (a), (b), and

(c) $\begin{cases} b_{102} = c_{20}b_{021} + c_{21}b_{111} \\ \qquad + c_{22}(c_{20}a_{201} + c_{21}a_{111} + c_{22}a_{102}) \\ b_{012} = c_{20}b_{111} + c_{21}b_{201} \\ \qquad + c_{22}(c_{20}a_{111} + c_{21}a_{021} + c_{22}a_{012}) \end{cases}$

are satisfied. Again the geometry in $\mathbf{R}^3$ is quite nice. For C^1, conditions (a) and (b) mean that the following three pairs of triangles are coplanar in $\mathbf{R}^3$:

$\{a_{300}, a_{210}, a_{201}\}$ and
$\{b_{300}, b_{210}, b_{201}\}$;

$\{a_{210}, a_{120}, a_{111}\}$ and
$\{b_{210}, b_{120}, b_{111}\}$;

$$\{a_{120}, a_{030}, a_{021}\} \quad \text{and}$$
$$\{b_{120}, b_{030}, b_{021}\}.$$

See Figure 5.7. For C^2, see the geometric interpretation in Figure 5.8.

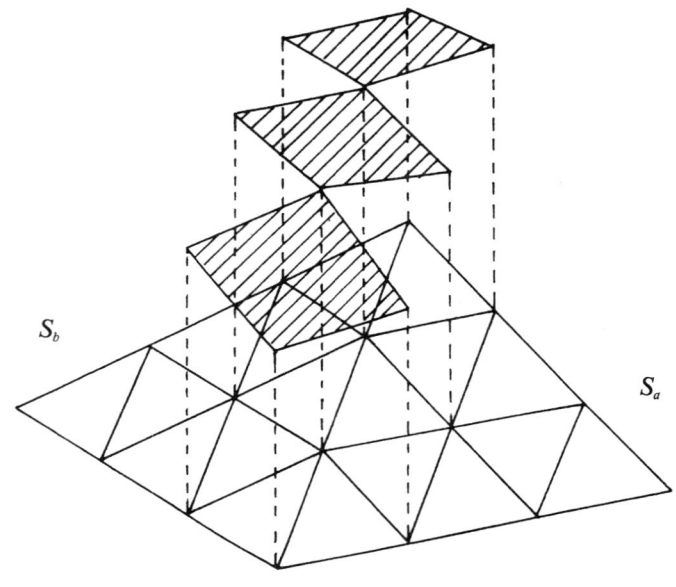

FIG. 5.7

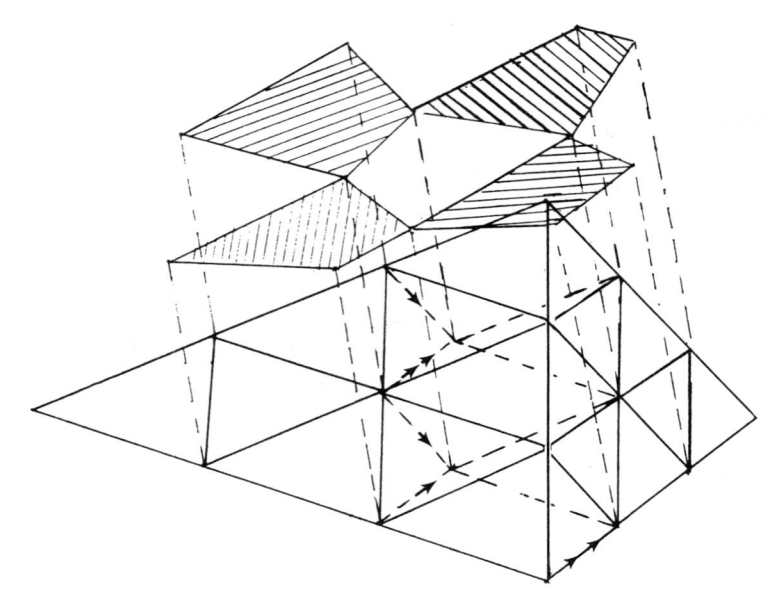

FIG. 5.8

Let us consider two important situations in the bivariate setting with $S_a \cap S_b = T_1 = \langle \mathbf{x}^0, \mathbf{x}^1 \rangle$. First suppose that $S_a \cup S_b$ is a parallelogram (cf. Figure 5.9). Then it is clear that

$$c_0 = c_1 = -c_2 = 1,$$

so that (5.8) becomes

$$\Delta_{20}^\gamma b_{ij0}^n = (\Delta_{10} - \Delta_{20})^\gamma a_{ij0}^n.$$

Upon simplification, we have the following result.

COROLLARY 5.1. *Let* $S_a \cap S_b = T_1 = \langle \mathbf{x}^0, \mathbf{x}^1 \rangle$ *and* $S_a \cup S_b$ *be a parallelogram. Then* $F \in C^r(S_a \cup S_b)$ *if and only if*

$$(5.9) \qquad b_{ijk}^n = \sum_{\ell=0}^{k} (-1)^\ell \binom{k}{\ell} \sum_{m=0}^{k-\ell} \binom{k-\ell}{m} a_{i+k-\ell-m,j+m,\ell}^n$$

for all $i + j = n - k$ *and* $k = 0, \cdots, r$.

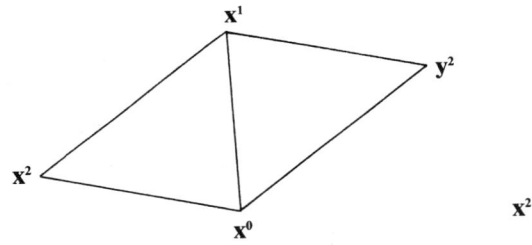

FIG. 5.9 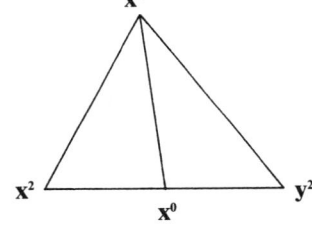   FIG. 5.10

The other important situation in the bivariate setting is when $\mathbf{x}^0, \mathbf{x}^2$, and $\mathbf{y}^2$ become colinear with $\mathbf{x}^0$ equidistant from $\mathbf{x}^2$ and $\mathbf{y}^2$ as shown in Figure 5.10. Then it follows that

$$c_0 = 2, \quad c_1 = 0, \quad c_2 = -1,$$

so that (5.8) becomes

$$\Delta_{20}^\gamma b_{ij0}^n = (-1)^\gamma \Delta_{20}^\gamma a_{ij0}^n.$$

Upon simplification, we have the following result.

COROLLARY 5.2. *Let* $S_a \cap S_b = \langle \mathbf{x}^0, \mathbf{x}^1 \rangle$ *such that* $S_a \cup S_b$ *is a triangle* $\langle \mathbf{x}^1, \mathbf{x}^2, \mathbf{y}^2 \rangle$ *and* $\mathbf{x}^0$ *is the mid-point of* $\langle \mathbf{x}^2, \mathbf{y}^2 \rangle$ *as shown in Figure* 5.10. *Then* $f \in C^r(S_a \cup S_b)$ *if and only if*

$$(5.10) \qquad b_{ijk}^n = \sum_{\ell=0}^{k} (-1)^\ell 2^{k-\ell} \binom{k}{\ell} a_{i+k-\ell,j,\ell}^n$$

for all $i + j = n - k$ *and* $k = 0, \cdots, r$.

5.3. Smoothness conditions for adjacent parallelepipeds. For simplicity, we only consider the case $\alpha = (n, \cdots, n)$ where $n \in \mathbf{Z}_+$. Let $\widetilde{T}_k = \langle \mathbf{u}^1, \cdots, \mathbf{u}^{2^k} \rangle$ be a k-parallelepiped in $\mathbf{R}^s$, $1 \le k \le s-1$, and let

$$\widetilde{S}_a = \langle \mathbf{u}^1, \cdots, \mathbf{u}^{2^k}, \mathbf{u}^{2^k+1}, \cdots, \mathbf{u}^{2^s} \rangle$$
$$\widetilde{S}_b = \langle \mathbf{u}^1, \cdots, \mathbf{u}^{2^k}, \mathbf{w}^{2^k+1}, \cdots, \mathbf{w}^{2^s} \rangle$$

be two adjacent s-parallelepipeds with $\widetilde{T}_k = \widetilde{S}_a \cap \widetilde{S}_b$. Suppose that $G(\mathbf{x}) \in C(\widetilde{S}_a \cup \widetilde{S}_b)$ with

$$G(\mathbf{x})|_{\widetilde{S}_a} = \widetilde{P}_n(\mathbf{x}) = \sum_{\gamma \le (n,\cdots,n)} \tilde{a}_\gamma \tilde{\phi}_\gamma^n(\mu)$$
$$G(\mathbf{x})|_{\widetilde{S}_b} = \widetilde{Q}_n(\mathbf{x}) = \sum_{\gamma \le (n,\cdots,n)} \tilde{b}_\gamma \tilde{\phi}_\gamma^n(\nu)$$

where $\mu = (\mu_1, \cdots, \mu_s)$ and $\nu = (\nu_1, \cdots, \nu_s)$ are the barycentric coordinates of $\mathbf{x}$ relative to $\widetilde{S}_a$ and $\widetilde{S}_b$, respectively. We arrange the coordinates so that $\mathbf{x} \in \widetilde{T}_k$ if and only if $\mu_{k+1}(\mathbf{x}) = \cdots = \mu_s(\mathbf{x}) = \nu_{k+1}(\mathbf{x}) = \cdots \nu_s(\mathbf{x}) = 0$ and $\mu_i(\mathbf{x}) = \nu_i(\mathbf{x})$ for $i = 1, \cdots, k$. In addition, we also arrange the points $\mathbf{u}^i, \mathbf{w}^i$, so that $\mathbf{u}^1$ is the *reference point* in the sense that $\mu_i(\mathbf{u}^1) = 0 = \nu_i(\mathbf{u}^1), i = 1, \cdots, s, \mu_j(\mathbf{u}^{j+1}) = 1 = \nu_j(\mathbf{u}^{j+1}), j = 1, \cdots, k$, and $\mu_{k+\ell}(\mathbf{u}^{2^k+\ell}) = 1 = \nu_{k+\ell}(\mathbf{w}^{2^k+\ell}), \ell = 1, \cdots, s-k$.

In the following we will use the notation:

$$\widetilde{D}_j f = D_{\mathbf{w}^{2^k+j-k} - \mathbf{u}^1} f,$$
$$\widehat{D}_j f = D_{\mathbf{u}^{2^k+j-k} - \mathbf{u}^1} f,$$
$$\triangle_j a_\gamma = a_{s_j \gamma} - a_\gamma.$$

Note that for $j = 1, \cdots, k$, we have

$$\widetilde{D}_j f = \widehat{D}_j f = D_{\mathbf{u}^{j+1} - \mathbf{u}^1} f.$$

We have the following result.

LEMMA 5.2. *For $j = 1, \cdots, s$, if $n_j = (n, \cdots, n, n-1, n, \cdots, n)$ where $n-1$ is at the jth component, then*

$$(\widetilde{D}_j \widetilde{P}_n)(\mathbf{x}) = n \sum_{\gamma \le n_j} \triangle_j \tilde{a}_\gamma \tilde{\phi}_\gamma^{n_j}(\mu_1(\mathbf{x}), \cdots, \mu_s(\mathbf{x})).$$

To study smoothness conditions, we must introduce the notation for degree raising. This is one of the disadvantages in working with coordinate degrees. Let

$$E_j \tilde{a}_\gamma^\sigma = \frac{\gamma_j}{\sigma_j+1} \tilde{a}_{s_j^{-1}\gamma}^\sigma + \left(1 - \frac{\gamma_j}{\sigma_j+1}\right) \tilde{a}_\gamma^\sigma$$

where $\sigma = (\sigma_1, \cdots, \sigma_s), \gamma = (\gamma_1, \cdots, \gamma_s)$, and

$$s_j^{-1}\gamma = (\gamma_1, \cdots, \gamma_{j-1}, \gamma_j - 1, \gamma_{j+1}, \cdots, \gamma_s),$$

and set $\tilde{a}_\gamma^{s_j\sigma} = E_j \tilde{a}_\gamma^\sigma$. The following result is a consequence of the above lemma. The proof can be found in the detailed version of Chui and Lai [66].

THEOREM 5.2. *Let* $\mathbf{c}^j = (c_{j1}, \cdots, c_{js})$, *where* $j = k+1, \cdots, s$, *be chosen to satisfy*

$$\mathbf{w}^{2^k+j-k} - \mathbf{u}^1 = \sum_{i=1}^{k} c_{ji}(\mathbf{u}^{i+1} - \mathbf{u}^1)$$

$$+ \sum_{i=k+1}^{s} c_{ji}(\mathbf{u}^{2^k+i-k} - \mathbf{u}^1).$$

Then for each $r \in \mathbf{Z}_+$, $G \in C^r(\widetilde{S}_a \cup \widetilde{S}_b)$ *if and only if*

$$\Delta_{k+1}^{\beta_{k+1}} \cdots \Delta_s^{\beta_s} \tilde{b}_\gamma = \sum_{|\alpha|=|\beta|} \left(\sum_{\substack{\eta^{k+1}+\cdots+\eta^s=\alpha \\ |\eta^j|=\beta_j, k+1 \leq j \leq s}} \prod_{j=k+1}^{s} \frac{\beta_j!(\mathbf{c}^j)^{\eta^j}}{\eta^j!} \right)$$

$$\times \left(\frac{(n!)^k (n-\beta_{k+1})! \cdots (n-\beta_s)!}{(n-\alpha_1)! \cdots (n-\alpha_s)!} \right)$$

$$\times \Delta_1^{\alpha_1} \cdots \Delta_s^{\alpha_1} E_1^{\alpha_1} \cdots E_k^{\alpha_k} \tilde{a}_\gamma^{(n-\alpha_1, \cdots, n-\alpha_k, n, \cdots, n)}$$

for all $\beta = (0, \cdots, 0, \beta_{k+1}, \cdots, \beta_s)$ *with* $|\beta| \leq r$ *and* $\gamma = (\gamma_1, \cdots, \gamma_k, 0, \cdots, 0)$ *with* $\gamma_1, \cdots, \gamma_k \leq n$, *where* $\tilde{a}_\gamma^{(n,\cdots,n)} = \tilde{a}_\gamma$.

5.4. Smoothness conditions for mixed partitions. This is a more complicated situation. We will only consider the bivariate setting. Let $T = \langle \mathbf{u}^0, \mathbf{u}^1 \rangle$, $S_a = \langle \mathbf{u}^0, \mathbf{u}^1, \mathbf{u}^2 \rangle$, and $\widetilde{S}_b = \langle \mathbf{u}^0, \mathbf{u}^1, \mathbf{w}^2, \mathbf{w}^3 \rangle$ with $S_a \cap \widetilde{S}_b = T$. See Figure 5.11, where S_a is a triangle and $\widetilde{S}_b$ a parallelogram.

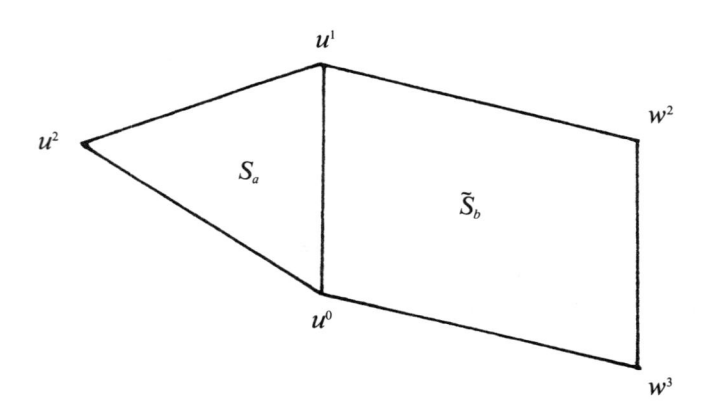

FIG. 5.11

Consider $H(\mathbf{x}) \in C(S_a \cup \widetilde{S}_b)$ with

$$H(\mathbf{x})|_{S_a} = \sum_{|\alpha|=n} a_\alpha^n \phi_\alpha^n(\lambda_0(\mathbf{x}), \lambda_1(\mathbf{x}), \lambda_2(\mathbf{x}))$$

and

$$H(\mathbf{x})|_{\widetilde{S}_b} = \sum_{\beta \leq (n,n)} \tilde{b}_\beta^n \tilde{\phi}_\beta^n(\mu_1(\mathbf{x}), \mu_2(\mathbf{x})),$$

where $\mathbf{x} \in T$ if and only if $\lambda_2(\mathbf{x}) = \mu_2(\mathbf{x}) = 0$, and we have set $\lambda_1(\mathbf{u}^0) = \mu_1(\mathbf{u}^0) = 0$, and $\lambda_1(\mathbf{u}^1) = \mu(\mathbf{u}^1) = 1$.

In the following, we will use the notation $\alpha = (i, j, k)$ and

$$Ra_\alpha^m = \frac{j}{m+1} a_{i+1,j-1,k}^m + \left(1 - \frac{j}{m+1}\right) a_{ijk}^m = a_\alpha^{m+1}.$$

Also, let p and q be determined by the equation

$$\mathbf{w}^3 - \mathbf{u}^0 = p(\mathbf{u}^1 - \mathbf{u}^0) + q(\mathbf{u}^2 - \mathbf{u}^0).$$

We have the following result.

THEOREM 5.3. *For each $r \in \mathbf{Z}_+$, $H(\mathbf{x})$ is in $C^r(S_a \cup \widetilde{S}_b)$ if and only if*

$$\Delta_2^k \tilde{b}_{j0} = R^k (p\Delta_{10} + q\Delta_{20})^k a_{n-j-k,j,0}^{n-k}$$

for all $j = 0, \cdots, n-k$ and $0 \leq k \leq r$.

More details can be found in Chui and Lai [66].

CHAPTER 6

Finite Elements and Vertex Splines

As we recall from Chapter 1, in univariate spline theory for a certain prescribed smoothness requirement, say C^r where $r \in \mathbf{Z}_+$, and a given partition (or knot sequence), the lowest degree of piecewise polynomials is $r + 1$, and a basis of the space of these (spline) functions is given by functions with minimal supports, called *B-splines*. We have also seen, however, that any multivariate extension would have to be much more complicated. An important problem is that for the prescribed smoothness $C^r, r \in \mathbf{Z}_+$, and a given grid partition $\triangle$, determine all locally supported piecewise polynomial functions in C^r on the grid $\triangle$ with *lowest* degree. While we still cannot achieve this goal, we will be satisfied in this chapter with determining all locally supported piecewise polynomial functions in C^r on $\triangle$ whose supports contain *at most one* grid point in the interior. These functions are called *vertex splines* first introduced in Chui and Lai [64] in the bivariate setting, and Chui and Lai [66] in the more general setting.

6.1. Vertex splines. We will use the notations introduced in Chapter 5. In particular, a simplex S_k with $k+1$ vertices $\mathbf{x}^0, \cdots, \mathbf{x}^k$ in $\mathbf{R}^s$ is denoted by $S_k = \langle \mathbf{x}^0, \cdots, \mathbf{x}^k \rangle$ where $k \leq s$; and for positive k, it is called a *k-simplex* if

$$\operatorname{vol}_k S_k > 0.$$

For $k = 0$, S_0 is a point, and we will also call S_0 a *zero-simplex* for convenience. If $S = \langle \mathbf{x}^0, \cdots, \mathbf{x}^s \rangle$ is an s-simplex, then for each k, $0 \leq k < s$, a k-simplex

$$\langle \mathbf{x}^{i_0}, \cdots, \mathbf{x}^{i_k} \rangle,$$

where $0 \leq i_0 < \cdots < i_k \leq s$, is also called a *k-facet* of S.

We first generalize the notion of (regular) triangulation discussed in §4.3 to a simplicial partition of a (closed) region Ω in $\mathbf{R}^s$ as follows: A finite collection of s-simplices $\{\Omega_j\}$ in $\mathbf{R}^s$ is called a *simplicial partition* of Ω if

(i) int $(\Omega_i) \cap$ int $(\Omega_j) = \emptyset$ for $i \neq j$,

and

(ii) either $\Omega_i \cap \Omega_j = \emptyset$, or $\Omega_i \cap \Omega_j$ is a k-simplex which is a common k-facet of Ω_i and Ω_j, where $0 \leq k < s$ and $i \neq j$.

Similarly, a parallelepiped with 2^k vertices and positive k-dimensional volume is called a *k-parallelepiped*, and again for convenience, for $k = 0$, a point is also called a *zero-parallelepiped*. Let

$$\widetilde{S} = \langle \mathbf{w}^1, \cdots, \mathbf{w}^{2^s} \rangle$$

be an s-parallelepiped. Then an $(s-1)$-parallelepiped

$$\langle \mathbf{w}^{i_1}, \cdots, \mathbf{w}^{i_{2^{s-1}}} \rangle,$$

where $0 \leq i_1 < \cdots < i_{2^{s-1}} \leq 2^s$, is called an $(s-1)$-*facet* of $\widetilde{S}$ if it is a subset of the boundary of $\widetilde{S}$. For $k = s-2, \cdots, 0$, we can now define inductively a k-*facet* of $\widetilde{S}$ as a k-parallelepiped which is a subset of the boundary of some $(k+1)$-facet of $\widetilde{S}$.

Analogous to a simplicial partition, we may define a *parallelepiped partition* if (i) and (ii) above are satisfied, where in statement (ii), the appropriate replacement of k-simplex by k-parallelepiped is necessary.

In applications such as numerical grid generation, it is more economical to consider a mixed partition. For instance, in $\mathbf{R}^2$, we may mix triangles with parallelograms. A typical situation is to use parallelograms in the interior and triangles near the boundary as in Figure 6.1.

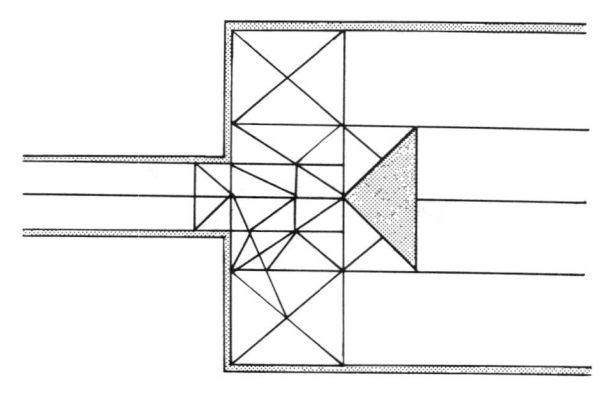

FIG. 6.1

In the following, if $s = 2$, we may consider a mixed partition as we just discussed such that (i) and (ii) are satisfied, but for $s > 2$, we will be concerned only with either a simplicial partition or a parallelepiped partition. Let Δ denote any of these partitions of $\Omega \subset \mathbf{R}^s$, and we will use the same notation $S_d^r(\Delta)$ for functions in $C^r(\Omega)$ whose restrictions on each Ω_i is a polynomial in π_d^s. Of course, this is called a multivariate spline space with grid partition Δ. We will also consider the subspace $\widehat{S}_d^r(\Delta)$ of *super splines* of $S_d^r(\Delta)$, consisting of $f \in S_d^r(\Delta)$ such that f is of class $C^{2^{s-k-1}r}$ at every k-dimensional facet of the partition, where $0 \leq k < s$.

We are ready to give a precise definition of vertex splines:

A *k-vertex spline* $V(\mathbf{x})$, $0 \leq k \leq s$, is a locally supported function in $\widehat{S}_d^r(\Delta)$ whose support contains one (and only one) k-simplex or k-parallelepiped which is the common k-facet of all cells (i.e., s-simplices or s-parallelepipeds) of Δ that are contained in the support of $V(\mathbf{x})$.

The collection of all k-vertex splines, $0 \leq k \leq s$, will simply be called the set of all *vertex splines*. Of course, all vertex splines are super splines. The following result can be found in Chui and Lai [66].

THEOREM 6.1. *Let Δ be either a simplicial or a parallelepiped partition and $\widehat{S}_d^r(\Delta)$ the subspace of all super splines, where $d \geq 2^s r + 1$. Then the collection of all vertex splines spans $\widehat{S}_d^r(\Delta)$.*

We believe that the same result still holds for mixed partitions with both simplices, parallelepipeds, etc., although it still requires some work to establish this claim. It is essentially proved in Ženiček [206], [207] for $s = 2$ and Le Méhaute [146], [147] for $s > 2$ that for a simplicial partition Δ, the super spline subspace $\widehat{S}_d^r(\Delta)$ provides the full approximation order if $d \geq 2^s r + 1$. The same result also holds for a parallelepiped partition Δ (cf. Chui and Lai [66]). Hence, vertex splines are very useful for constructing approximation and interpolation schemes, as we will see in §6.4, for instance. However, the lower bound $2^s r + 1$ is not sharp. For $s = 2$, the sharp lower bound is $3r + 2$ as discussed in §4.4.

6.2. Generalized vertex splines. In Chui and Lai [66], it is shown that vertex splines exist provided $d \geq 2^s r + 1$. Since high degree polynomials are usually undesirable due to various reasons such as large local dimension and difficulty in shape control, we would like to reduce d as much as possible for a given r. One way to accomplish this is to refine the grid partition. In finite element methods (FEM), the element obtained in this manner is called a macro element. We recommend such a procedure only if it is done in a very symmetric way and the dimension is not increased. This leads to our notion of *generalized vertex splines*. Hence, the support of a generalized vertex spline is the same as that of one of the k-vertex splines, $0 \leq k \leq s$, and the only differences are that each cell (i.e., simplex or parallelepiped) in the support is refined in a symmetric way so as to reduce the degree d but still satisfy the smoothness condition C^r, and that the super smoothness condition $C^{2^{s-k-1}r}$ is dropped. For instance, in Figure 6.2 we give the supports of all k-vertex splines in $\mathbf{R}^2$, $k = 0, 1, 2$ and refine them to give the supports of some generalized vertex splines by drawing in all medians of each triangle.

In Chui and He [60], bivariate C^1 quadratic generalized vertex splines on an arbitrary (regular) triangulation $\triangle$ are constructed and their properties are studied. Following an idea of Powell and Sabin [164] and Heindl [131] we divide each triangular cell Ω_j in $\triangle$ into 12 subtriangles using the three medians as shown in Figure 6.3.

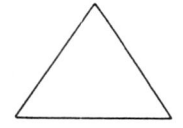

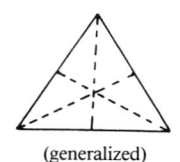

support of 2-vertex spline   (generalized)

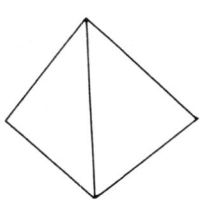

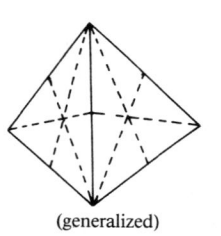

support of 1-vertex spline   (generalized)

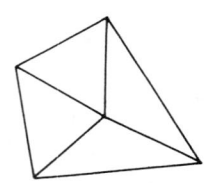

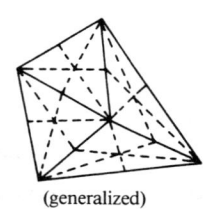

support of zero-vertex spline   (generalized)

FIG. 6.2

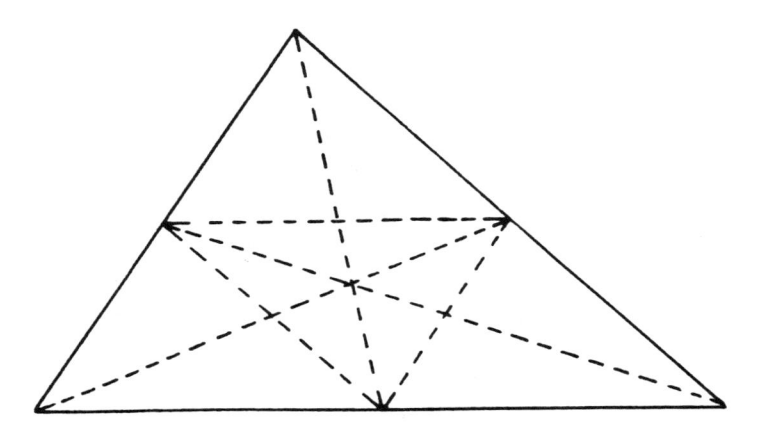

FIG. 6.3

Let us call the refinement $\widehat{\triangle}$, so that $\widehat{\triangle} \supset \triangle$. Let $S_2^1(\widehat{\triangle})$ be the bivariate C^1 quadratic spline space on $\widehat{\triangle}$. The dimension of $S_2^1(\widehat{\triangle})$ can be determined as follows: Consider any triangular cell Ω_j of the original triangulation $\triangle$ and treat it as a region to be partitioned by the refinement as in Figure 6.3 above. Since this is a crosscut partition, its dimension is

$$\binom{2+2}{2} + 6\binom{2-1+1}{2} + C_2^1(3) = 6 + 6 + 0 = 12$$

from Theorem 4.4. Hence, by assigning three values $(\delta, \delta_x, \delta_y)$ at each vertex and one value δ_n at the mid-point of each edge, giving a total of 12 conditions, there is one and only one function in this space, i.e., $S_2^1(\widehat{\triangle}, \Omega_j)$, that interpolates these 12 data values. Here, δ means function value, δ_x and δ_y represent values of the first partial derivatives with respect to x and y, and δ_n denotes value of the outer normal derivative. This is done in Powell and Sabin [164], and Heindl [131], and it is a standard technique in finite element methods for moving from one "element" to a neighboring one. Now, by using the $(\delta, \delta_x, \delta_y)$ and δ_n values on each triangular region of $\triangle$ to match the neighboring ones, we obtain the whole space $S_2^1(\widehat{\triangle})$. That is, the dimension of $S_2^1(\widehat{\triangle})$ is exactly

(6.1) \hspace{2em} 3(\#vertices) + \#edges

of $\triangle$, where both boundary and interior vertices and edges are counted.

The above argument also happens to give all the generalized vertex splines on $\triangle$. By taking all $(\delta, \delta_x, \delta_y)$ values and all except one δ_n values to be zero, we have the generalized 1-vertex splines with supports shown in Figure 6.4.

76   CHAPTER 6

(interior)   (boundary)
generalized 1-vertex spline   generalized 1-vertex spline

FIG. 6.4

Similarly, by taking all $(\delta, \delta_x, \delta_y)$ except one triple, say the one at vertex A, to be zero, and all δ_n values, except the edges connecting to A, to be zero, we have all the generalized zero-vertex splines with supports shown in Figure 6.5.

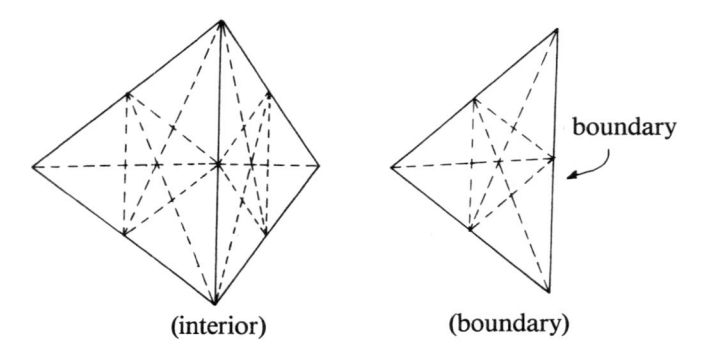

(interior)   (boundary)
generalized zero-vertex spline   generalized zero-vertex spline

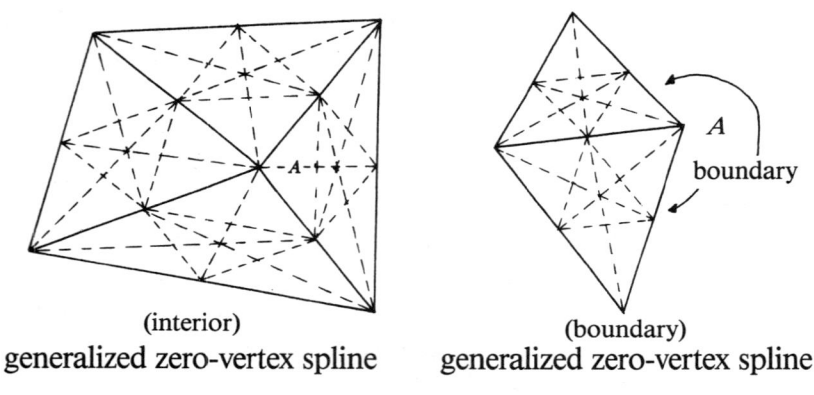

FIG. 6.5

In fact, the Bézier nets of all the above generalized vertex splines have been calculated in Chui and He [60] (see Example 6.2 in §6.3 to follow). It is clear that these functions form a basis of $S_2^1(\widehat{\triangle})$ where $\widehat{\triangle} \supset \triangle$.

The first question we may ask is what δ_n values should be chosen on the edges connected to A for a generalized zero-vertex spline with vertex A. To answer this question, let us label all vertices of the triangulation $\triangle$ by $A_1, \cdots, A_V$ where Ball interior and boundary vertices are included. To each vertex $A_i, i = 1, \cdots, V$, we consider three generalized zero-vertex splines

(6.2) $S_i, \quad T_i, \quad U_i$

whose $(\delta, \delta_x, \delta_y)$ values at A_i are $(1,0,0)$, $(0,1,0)$ and $(0,0,1)$, respectively. Let the coordinates of A_i be $A_i = (a_1^i, a_2^i)$. Unfortunately, we have to relabel all the vertices as follows. Let all the vertices that share common edges with A_i be labeled, in the counterclockwise direction, by $B_{i1}, \cdots, B_{in_i}$ as in Figure 6.6 and set $B_{ij} = (b_1^{ij}, b_2^{ij})$. The normal derivatives (or δ_n values) at the center of the edge joining B_{ij} to A_i for the spline functions S_i, T_i, U_i will be denoted by s_{ij}, t_{ij}, u_{ij}, respectively. The following result is established in Chui and He [60].

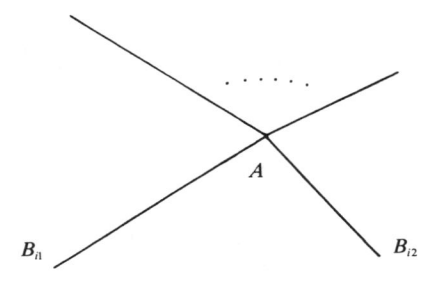

FIG. 6.6

THEOREM 6.2. *A necessary and sufficient condition for*

$$(6.3) \quad \sum_{i=1}^{V} \left[f(A_i) S_i(x,y) + \frac{\partial}{\partial x} f(A_i) T_i(x,y) + \frac{\partial}{\partial y} f(A_i) U_i(x,y) \right]$$
$$= f(x,y)$$

for all $f \in \pi_2^2$ and $(x,y) \in \Omega$ is that

(i) $s_{ij} = 0$,

(ii) $t_{ij} = \frac{a_2^i - b_2^{ij}}{2((a_1^i - b_1^{ij})^2 + (a_2^i - b_2^{ij})^2)^{1/2}}$,

and

(iii) $u_{ij} = \frac{b_1^{ij} - a_1^i}{2((a_1^i - b_1^{ij})^2 + (a_2^i - b_2^{ij})^2)^{1/2}}$,

for all $j = 1, \cdots, n_i$ and $i = 1, \cdots, V$.

In other words, by using the δ_n values given by (i) - (iii) above, we may drop *all* the (generalized) 1-vertex splines and still obtain the full approximation order of 3. We will denote by S_i^*, T_i^*, and U_i^* the generalized zero-vertex splines with the normal derivative values given by (i), (ii), and (iii), respectively.

Another important question is whether or not we could simplify the refinement from 12 subtriangles as in Figure 6.3 to 6 triangles as in Figure

6.2. The answer to this question is that the original triangulation must essentially be a three-directional mesh. We are not going into details here, but refer the reader to [51].

6.3. Polynomial interpolation formulas and examples. In constructing both vertex splines and generalized vertex splines, the smoothing formulas derived in Chapter 5 are essential. However, since vertex splines are constructed using Hermite interpolating conditions at the vertices, formulas for such interpolation purposes in terms of the Bézier nets are also important tools. A discussion of such formulas derived in a detailed version of Chui and Lai [66] would have to be very involved. We only mention here the following Taylor formula given in Chui and Lai [64] for an s-simplex. The notations are consistent with the ones in Chapter 5.

THEOREM 6.3. *Let $\langle \mathbf{x}^0, \cdots, \mathbf{x}^s \rangle$ be an s-simplex. Then the Taylor polynomial of a function $f(\mathbf{x})$ at the vertex $\mathbf{x}^0$ in Bézier representation with respect to this simplex is given by*

$$(6.4) \qquad (P_n f)(\mathbf{x}) =$$

$$\sum_{|\alpha|=n} \sum_{\substack{\beta_i \leq \alpha_i \\ i=1,\cdots,s}} \binom{\alpha_1}{\beta_1} \cdots \binom{\alpha_s}{\beta_s} \frac{(n-|\beta|)!}{n!} D_{10}^{\beta_1} \cdots D_{s0}^{\beta_s} f(\mathbf{x}^0) \phi_\alpha^n(\lambda(\mathbf{x}))$$

where $\alpha = (\alpha_0, \cdots, \alpha_s) \in \mathbf{Z}_+^{s+1}$.

This result can be proved by mathematical induction on the degree n.

In the following, we give two examples for the bivariate setting, one on vertex splines and the other on generalized vertex splines. Both are computed by using the formula (6.4). Let $\mathbf{x}^i$ be a vertex of the partition Δ which may be an interior or boundary vertex and let $\mathbf{x}^{ik}$ be the vertices that share common edges with $\mathbf{x}^i$ as in Figure 6.7.

As usual, write $\mathbf{x}^i = (x_1^i, x_2^i)$ and $\mathbf{x}^{ik} = (x_1^{ik}, x_2^{ik})$. In addition, denote by $\delta(\mathbf{x}^0, \mathbf{x}^1, \mathbf{x}^2)$ the signed area of the triangle $\langle \mathbf{x}^0, \mathbf{x}^1, \mathbf{x}^2 \rangle$; namely,

$$\delta(\mathbf{x}^0, \mathbf{x}^1, \mathbf{x}^2) = \frac{1}{2} \begin{vmatrix} 1 & x_1^0 & x_2^0 \\ 1 & x_1^1 & x_2^1 \\ 1 & x_1^2 & x_2^2 \end{vmatrix}.$$

FINITE ELEMENTS AND VERTEX SPLINES

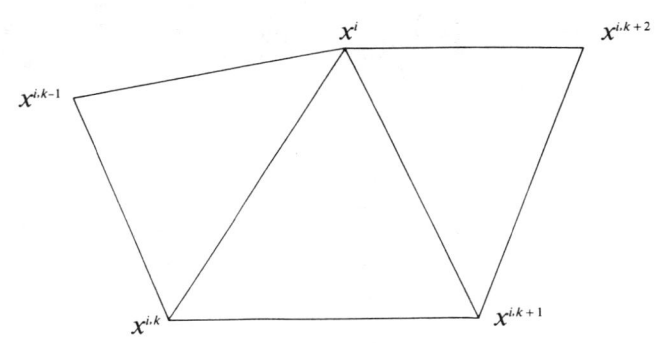

FIG. 6.7

Example 6.1. In this example, we give the vertex spline basis of the super spline subspace $\widehat{S}_5^1(\Delta)$. To each (interior or boundary) vertex, there are six zero-vertex splines which will be determined by the $(\delta, \delta_x, \delta_y, \delta_{xx}, \delta_{xy}, \delta_{yy})$ values. Here, δ_{xx}, δ_{xy}, and δ_{yy} denote values of the second partial derivatives. In addition, to each edge there is one 1-vertex spline. See Figures 6.8–6.14. We set

$$a_{k1} = \frac{\delta(\mathbf{x}^{i,k}, \mathbf{x}^{i,k+1}, \mathbf{x}^{i,k-1})}{\delta(\mathbf{x}^{i,k+1}, \mathbf{x}^i, \mathbf{x}^{i,k-1}) + \delta(\mathbf{x}^{i,k}, \mathbf{x}^{i,k+1}, \mathbf{x}^{i,k-1})},$$

$$a_{k2} = \frac{\delta(\mathbf{x}^{i,k+1}, \mathbf{x}^{i,k+2}, \mathbf{x}^{i,k})}{\delta(\mathbf{x}^{i,k+2}, \mathbf{x}^i, \mathbf{x}^{i,k}) + \delta(\mathbf{x}^{i,k+1}, \mathbf{x}^{i,k+2}, \mathbf{x}^{i,k})},$$

$$b_k = \frac{1}{5}(x_1^{i,k} - x_1^i), \qquad c_k = \frac{1}{5}(x_2^{i,k} - x_2^i),$$

$$d_k = \frac{1}{20}(x_1^{i,k} - x_1^i)^2,$$

$$e_k = \frac{1}{20}(x_2^{i,k} - x_2^i)^2,$$

$$f_k = \frac{1}{20}(x_1^{i,k} - x_1^i)(x_1^{i,k+1} - x_1^i),$$

$$g_k = \frac{1}{10}(x_1^{i,k} - x_1^i)(x_2^{i,k} - x_2^i),$$

$$\tilde{g}_k = \frac{1}{20}[(x_1^{i,k+1} - x_1^i)(x_2^{i,k} - x_2^i) + (x_2^{i,k+1} - x_2^i)(x_1^{i,k} - x_1^i)],$$

$$h_k = \frac{1}{20}(x_2^{i,k} - x_2^i)(x_2^{i,k+1} - x_2^i),$$

and

$$\ell_k = \delta(\mathbf{x}^i, \mathbf{x}^{i,k-1}, \mathbf{x}^{i,k}).$$

80   CHAPTER 6

(i) $(\delta, \delta_x, \delta_y, \delta_{xx}, \delta_{xy}, \delta_{yy}) = (1, 0, 0, 0, 0, 0)$ at the vertex $\mathbf{x}^i$:

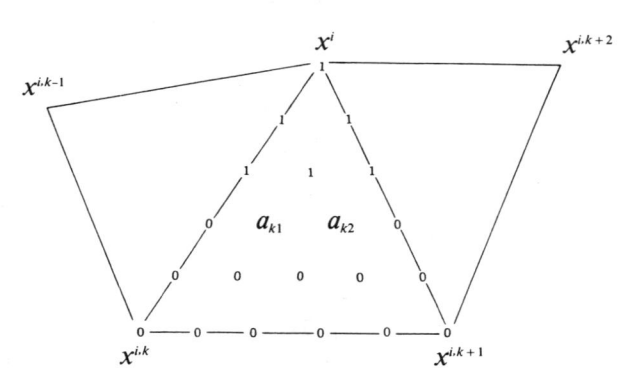

FIG. 6.8

(ii) $(\delta, \delta_x, \delta_y, \delta_{xx}, \delta_{xy}, \delta_{yy}) = (0, 1, 0, 0, 0, 0)$ at vertex $\mathbf{x}^i$:

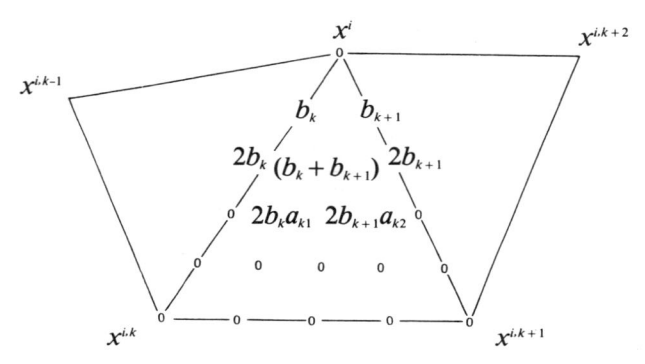

FIG. 6.9

(iii) $(\delta, \delta_x, \delta_y, \delta_{xx}, \delta_{xy}, \delta_{yy}) = (0, 0, 1, 0, 0, 0)$ at vertex $\mathbf{x}^i$:

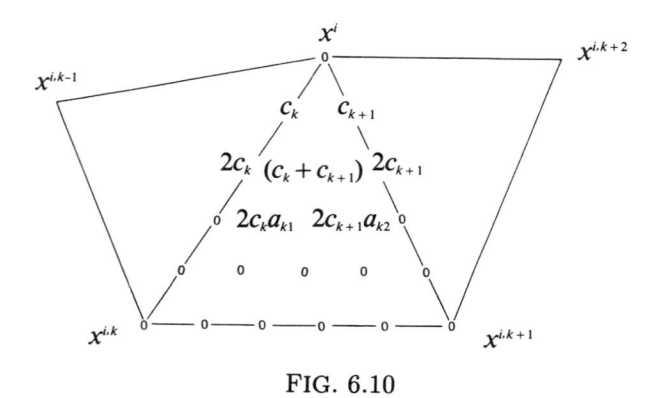

FIG. 6.10

FINITE ELEMENTS AND VERTEX SPLINES

(iv) $(\delta, \delta_x, \delta_y, \delta_{xx}, \delta_{xy}, \delta_{yy}) = (0, 0, 0, 1, 0, 0)$ at vertex $\mathbf{x}^i$:

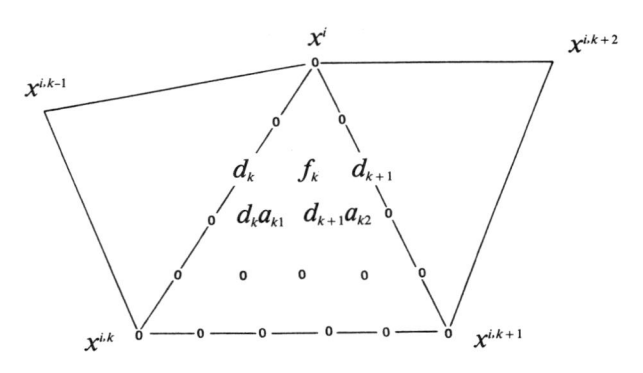

FIG. 6.11

(v) $(\delta, \delta_x, \delta_y, \delta_{xx}, \delta_{xy}, \delta_{yy}) = (0, 0, 0, 0, 1, 0)$ at vertex $\mathbf{x}^i$:

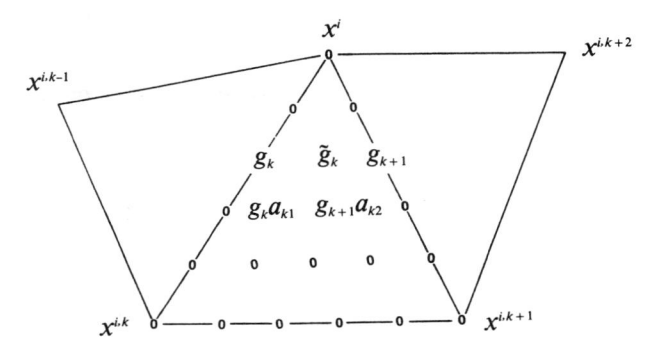

FIG. 6.12

(vi) $(\delta, \delta_x, \delta_y, \delta_{xx}, \delta_{xy}, \delta_{yy}) = (0, 0, 0, 0, 0, 1)$ at vertex $\mathbf{x}^i$:

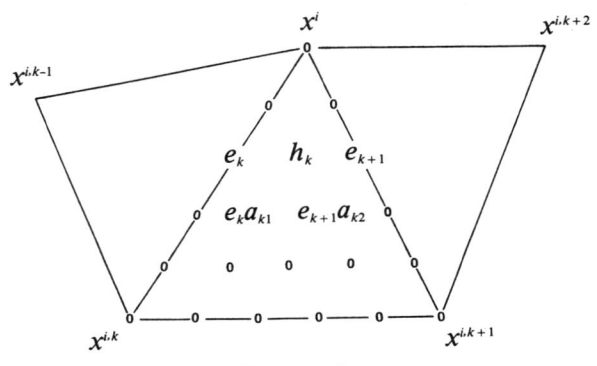

FIG. 6.13

(vii) 1-vertex spline:

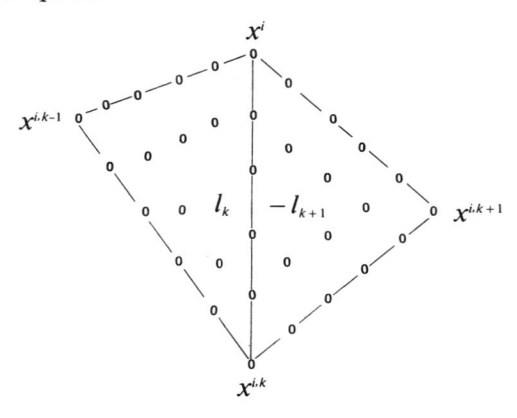

FIG. 6.14

Example 6.2. In this example, we give the generalized vertex spline basis of $S_2^1(\widehat{\Delta})$. To each (interior or boundary) vertex, there are three generalized zero-vertex splines which will be determined by the $(\delta, \delta_x, \delta_y)$ values, and to each edge there is one generalized 1-vertex spline.

(i) In Figure 6.15, we display S^* at vertex $\mathbf{x}^i$ where

$$\alpha_k = 1, \quad \alpha_{k+1} = 1,$$
$$\delta_k = 1/2, \quad \delta_{k+1} = 1/2, \quad \gamma_k = 1$$
$$\beta_{k1} = \frac{1}{2} - \frac{1}{4\ell_k^2}[(x_1^{i,k} - x_1^i)(x_1^{i,k+1} - x_1^{i,k})$$
$$+ (x_2^{i,k} - x_2^i)(x_2^{i,k+1} - x_2^{i,k})]$$
$$\beta_{k2} = \frac{1}{2} - \frac{1}{4\ell_{k+1}^2}[(x_1^{i,k+1} - x_1^i)(x_1^{i,k} - x_1^{i,k+1})$$
$$+ (x_2^{i,k+1} - x_2^i)(x_2^{i,k} - x_2^{i,k+1})]$$
$$\varsigma_{k1} = \frac{1}{2} - \frac{1}{6\ell_k^2}[(x_1^{i,k} - x_1^i)(2x_1^{i,k+1} - x_1^i - x_1^{i,k})$$
$$+ (x_2^{i,k} - x_2^i)(2x_2^{i,k+1} - x_2^i - x_2^{i,k})]$$
$$\varsigma_{k2} = \frac{1}{2} - \frac{1}{6\ell_{k+1}^2}[(x_1^{i,k+1} - x_1^i)(2x_1^{i,k} - x_1^i - x_1^{i,k+1})$$
$$+ (x_2^{i,k+1} - x_2^i)(2x_2^{i,k} - x_2^i - x_2^{i,k+1})$$
$$\eta_{k1} = \frac{1}{2} - \frac{1}{4\ell_k^2}[(x_1^{i,k} - x_1^i)(x_1^{i,k+1} - x_1^i)$$
$$+ (x_2^{i,k} - x_2^i)(x_2^{i,k+1} - x_2^i)]$$

$$\eta_{k2} = \frac{1}{2} - \frac{1}{4\ell_{k+1}^2}[(x_1^{i,k+1} - x_1^i)(x_1^{i,k} - x_1^i)$$
$$+ (x_2^{i,k+1} - x_2^i)(x_2^{i,k} - x_2^i)]$$
$$\ell_k^2 = (x_1^{i,k} - x_1^i)^2 + (x_2^{i,k} - x_2^i)^2$$
$$\ell_{k+1}^2 = (x_1^{i,k+1} - x_1^i)^2 + (x_2^{i,k+1} - x_2^i)^2$$
$$b_k = \frac{1}{2}(\beta_{k1} + \beta_{k2}) \qquad c_k = \frac{1}{2}(\varsigma_{k1} + \varsigma_{k2})$$

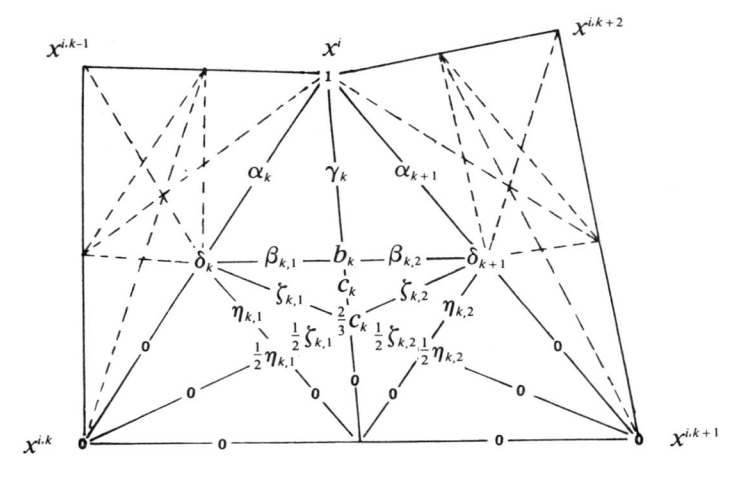

FIG. 6.15

(ii) In Figure 6.16, we display T^* at vertex $\mathbf{x}^i$ where

$$\alpha_k = \frac{1}{4}(x_1^{i,k} - x_1^i)$$
$$\alpha_{k+1} = \frac{1}{4}(x_1^{i,k+1} - x_1^i)$$
$$\delta_k = \frac{1}{8}(x_1^{i,k} - x_1^i)$$
$$\delta_{k+1} = \frac{1}{8}(x_1^{i,k+1} - x_1^i)$$
$$\gamma_k = \frac{1}{8}(x_1^{i,k} + x_1^{i,k+1} - 2x_1^i)$$
$$\beta_{k1} = \frac{1}{8}(x_1^{ik} - x_1^i) - \frac{1}{16\ell_k^2}[(x_1^{i,k} - x_1^i)^2(x_1^{i,k+1} - x_1^{i,k})$$
$$- (x_2^{i,k} - x_2^i)^2(x_1^{i,k+1} - x_1^{i,k})$$
$$+ 2(x_1^{i,k} - x_1^i)(x_2^{i,k} - x_2^i)(x_2^{i,k+1} - x_2^{i,k})]$$

$$\beta_{k2} = \frac{1}{8}(x_1^{i,k+1} - x_1^i) - \frac{1}{16\ell_{k+1}^2}[(x_1^{i,k+1} - x_1^i)^2(x_1^{i,k} - x_1^{i,k+1})$$
$$- (x_2^{i,k+1} - x_2^i)^2(x_1^{i,k} - x_1^{i,k+1})$$
$$+ 2(x_1^{i,k+1} - x_1^i)(x_2^{i,k+1} - x_2^i)(x_2^{i,k} - x_2^{i,k+1})]$$
$$\varsigma_{k1} = \frac{1}{8}(x_1^{i,k} - x_1^i) - \frac{1}{24\ell_k^2}[(x_1^{i,k} - x_1^i)^2(2x_1^{i,k+1} - x_1^i - x_1^{i,k})$$
$$- (x_2^{i,k} - x_2^i)^2(2x_1^{i,k+1} - x_1^i - x_1^{i,k})$$
$$+ 2(x_1^{i,k} - x_1^i)(x_2^{i,k} - x_2^i)(2x_2^{i,k+1} - x_2^i - x_2^{i,k})]$$
$$\varsigma_{k2} = \frac{1}{8}(x_1^{i,k+1} - x_1^i)$$
$$- \frac{1}{24\ell_{k+1}^2}[(x_1^{i,k+1} - x_1^i)^2(2x_1^{i,k} - x_1^i - x_1^{i,k+1})$$
$$- (x_2^{i,k+1} - x_2^i)^2(2x_1^{i,k} - x_1^i - x_1^{i,k+1})$$
$$+ 2(x_1^{i,k+1} - x_1^i)(x_2^{i,k+1} - x_2^i)(2x_2^{i,k} - x_2^i - x_2^{i,k+1})]$$
$$\eta_{k1} = \frac{1}{8}(x_1^{i,k} - x_1^i) - \frac{1}{16\ell_k^2}[(x_1^{i,k} - x_1^i)^2(x_1^{i,k+1} - x_1^i)$$
$$- (x_2^{i,k} - x_2^i)^2(x_1^{i,k+1} - x_1^i)$$
$$+ 2(x_1^{i,k} - x_1^i)(x_2^{i,k} - x_2^i)(x_2^{i,k+1} - x_2^i)]$$
$$\eta_{k2} = \frac{1}{8}(x_1^{i,k+1} - x_1^i) - \frac{1}{16\ell_{k+1}^2}[(x_1^{i,k+1} - x_1^i)^2(x_1^{i,k} - x_1^i)$$
$$- (x_2^{i,k+1} - x_2^i)^2(x_1^{i,k} - x_1^i)$$
$$+ 2(x_1^{i,k+1} - x_1^i)(x_2^{i,k+1} - x_2^i)(x_2^{i,k} - x_2^i)]$$
$$\ell_k^2 = (x_1^{i,k} - x_1^i)^2 + (x_2^{i,k} - x_2^i)^2$$
$$\ell_{k+1}^2 = (x_1^{i,k+1} - x_1^i)^2 + (x_2^{i,k+1} - x_2^i)^2$$
$$b_k = \frac{1}{2}(\beta_{k1} + \beta_{k2})$$
$$c_k = \frac{1}{2}(\varsigma_{k1} + \varsigma_{k2})$$

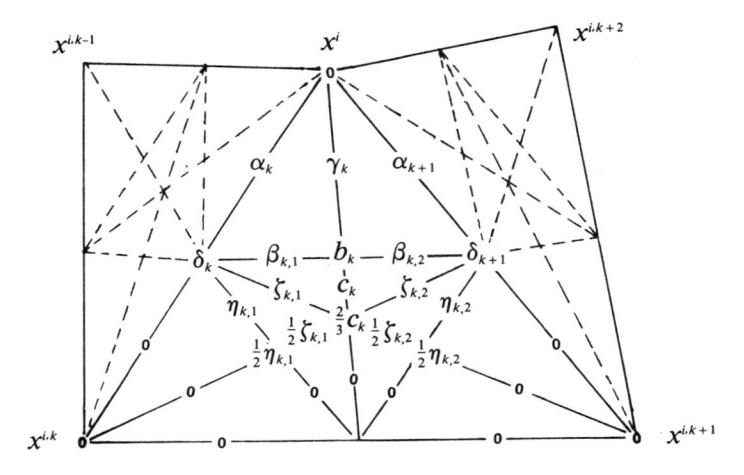

FIG. 6.16

(iii) In Figure 6.17, we display U^* at vertex $\mathbf{x}^i$ where

$$\alpha_k = \frac{1}{4}(x_2^{i,k} - x_2^i) \quad \alpha_{k+1} = \frac{1}{4}(x_2^{i,k+1} - x_2^i)$$

$$\delta_k = \frac{1}{8}(x_2^{i,k} - x_2^i) \quad \delta_{k+1} = \frac{1}{8}(x_2^{i,k+1} - x_2^i)$$

$$\gamma_k = \frac{1}{8}(x_2^{i,k} + x_2^{i,k+1} - 2x_2^i)$$

$$\beta_{k1} = \frac{1}{8}(x_2^{ik} - x_2^i) - \frac{1}{16\ell_k^2}[2(x_2^{i,k} - x_2^i)(x_1^{i,k} - x_1^i)$$
$$\times (x_1^{i,k+1} - x_1^{i,k}) + (x_2^{i,k} - x_2^i)^2(x_2^{i,k+1} - x_2^{i,k})$$
$$- (x_1^{i,k} - x_1^i)^2(x_2^{i,k+1} - x_2^{i,k})]$$

$$\beta_{k2} = \frac{1}{8}(x_2^{i,k+1} - x_2^i) - \frac{1}{16\ell_{k+1}^2}[2(x_2^{i,k+1} - x_2^i)(x_1^{i,k+1} - x_1^i)$$
$$\times (x_1^{i,k} - x_1^{i,k+1}) + (x_2^{i,k+1} - x_2^i)^2(x_2^{i,k} - x_2^{i,k+1})$$
$$- (x_1^{i,k+1} - x_1^i)^2(x_2^{i,k} - x_2^{i,k+1})]$$

$$\varsigma_{k1} = \frac{1}{8}(x_2^{i,k} - x_2^i)$$
$$- \frac{1}{24\ell_k^2}[2(x_2^{i,k} - x_2^i)(x_1^{i,k} - x_1^i)(2x_1^{i,k+1} - x_1^i - x_1^{i,k})$$
$$+ (x_2^{i,k} - x_2^i)^2(2x_2^{i,k+1} - x_2^i - x_2^{i,k})$$
$$- (x_1^{i,k} - x_1^i)^2(2x_2^{i,k+1} - x_2^i - x_2^{i,k})]$$

$$\varsigma_{k2} = \frac{1}{8}(x_2^{i,k+1} - x_2^i)$$
$$- \frac{1}{24\ell_{k+1}^2}[2(x_2^{i,k+1} - x_2^i)$$
$$\times (x_1^{i,k+1} - x_1^i)(2x_1^{i,k} - x_1^i - x_1^{i,k+1})$$
$$+ (x_2^{i,k+1} - x_2^i)^2(2x_2^{i,k} - x_2^i - x_2^{i,k+1})$$
$$- (x_1^{i,k+1} - x_1^i)^2(2x_2^{i,k} - x_2^i - x_2^{i,k+1})]$$
$$\eta_{k1} = \frac{1}{8}(x_2^{i,k} - x_2^i) - \frac{1}{16\ell_k^2}[2(x_2^{i,k} - x_2^i)(x_1^{i,k} - x_1^i)(x_1^{i,k+1} - x_1^i)$$
$$+ (x_2^{i,k} - x_2^i)^2(x_2^{i,k+1} - x_2^i) - (x_1^{i,k} - x_1^i)^2(x_2^{i,k+1} - x_2^i)]$$
$$\eta_{k2} = \frac{1}{8}(x_2^{i,k+1} - x_2^i) - \frac{1}{16\ell_{k+1}^2}[2(x_2^{i,k+1} - x_2^i)(x_1^{i,k+1} - x_1^i)$$
$$\times (x_1^{i,k} - x_1^i) + (x_2^{i,k+1} - x_2^i)^2(x_2^{i,k} - x_2^i)$$
$$- (x_1^{i,k+1} - x_1^i)^2(x_2^{i,k} - x_2^i)]$$
$$\ell_k^2 = (x_1^{i,k} - x_1^i)^2 + (x_2^{i,k} - x_2^i)^2$$
$$\ell_{k+1}^2 = (x_1^{i,k+1} - x_1^i)^2 + (x_2^{i,k+1} - x_2^i)^2$$
$$b_k = \frac{1}{2}(\beta_{k1} + \beta_{k2}) \qquad c_k = \frac{1}{2}(\varsigma_{k1} + \varsigma_{k2})$$

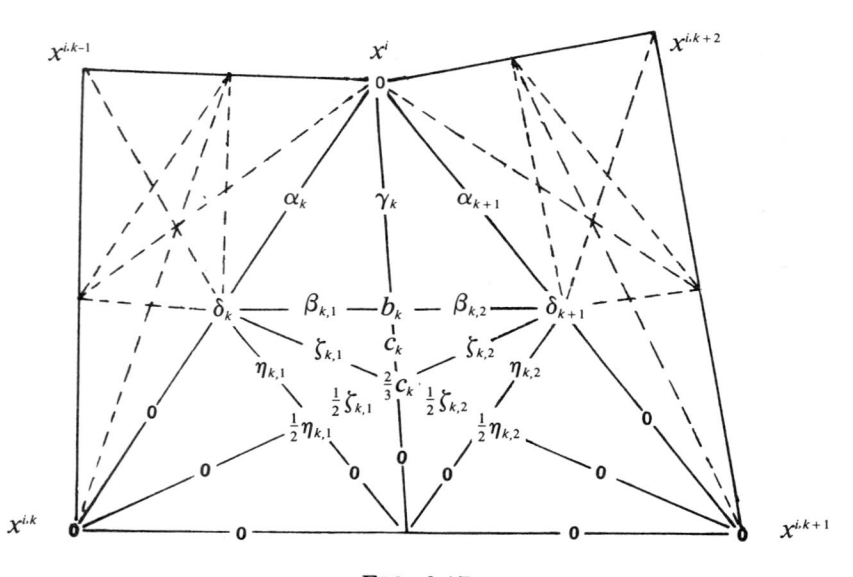

FIG. 6.17

FINITE ELEMENTS AND VERTEX SPLINES

(iv) In Figure 6.18, we display the generalized 1-vertex spline where

$$\beta_{k1} = \frac{1}{8\ell_k^2}[(x_1^{i,k} - x_1^i)(x_2^{i,k+1} - x_2^{i,k}) - (x_2^{i,k} - x_2^i)(x_1^{i,k+1} - x_1^{i,k})]$$

$$\beta_{k-1,2} = \frac{1}{8\ell_k^2}[(x_1^{i,k} - x_1^i)(x_2^{i,k-1} - x_2^{i,k}) - (x_2^{i,k} - x_2^i)(x_1^{i,k-1} - x_1^{i,k})]$$

$$\varsigma_{k1} = \frac{1}{12\ell_k^2}[(x_1^{i,k} - x_1^i)(2x_2^{i,k+1} - x_2^i - x_2^{i,k}) - (x_2^{i,k} - x_2^i)(2x_1^{i,k+1} - x_1^i - x_1^{i,k})]$$

$$\varsigma_{k-1,2} = \frac{1}{12\ell_k^2}[(x_1^{i,k} - x_1^i)(2x_2^{i,k-1} - x_2^i - x_2^{i,k}) - (x_2^{i,k} - x_2^i)(2x_1^{i,k-1} - x_1^i - x_1^{i,k})$$

$$\eta_{k1} = \frac{1}{8\ell_k^2}[(x_1^{i,k} - x_1^i)(x_2^{i,k+1} - x_2^i) - (x_2^{i,k} - x_2^i)(x_1^{i,k+1} - x_1^i)]$$

$$\eta_{k-1,2} = \frac{1}{8\ell_k^2}[(x_1^{i,k} - x_1^i)(x_2^{i,k-1} - x_2^i) - (x_2^{i,k} - x_2^i)(x_1^{i,k-1} - x_1^i)]$$

$$\ell_k^2 = (x_1^{i,k} - x_1^i)^2 + (x_2^{i,k} - x_2^i)^2$$

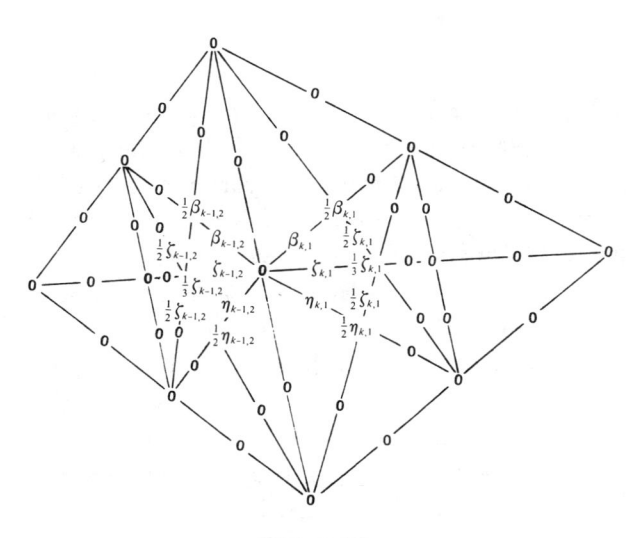

FIG. 6.18

6.4. Applications. Vertex splines or generalized vertex splines are useful in various problems in approximation and interpolation when a local basis would facilitate the computational scheme, the derivation of normal equations, and proof of certain results, etc. In addition to the trivial application to Hermite interpolation, it has applications to construction of a quasi-interpolation formula, least-squares approximation to both continuous and discrete data, shape-preserving approximation, CAGD, and reconstruction of a gradient field, etc. We will delay our discussion of the last three topics to Chapter 10 and only briefly study quasi-interpolation and least-square approximation here.

In many applications, only a set of discrete data that represents function values is given, and we are asked to construct an approximation formula that gives the full order of approximation. Since values of the derivatives are not known, we cannot use Hermite interpolation. We remark, however, that the problem is actually algebraic in nature in that we are supposed to determine a family of coefficients, say $\{a_{ij}^\alpha\}$, such that the "spline series"

$$(6.5) \qquad (Qf)(\mathbf{x}) = \sum_i \sum_\alpha \sum_{|j| \leq M} a_{ij}^\alpha f(\mathbf{u}^{i+j}) S_i^\alpha(\mathbf{x}),$$

for some $M > 0$, reproduces all polynomials $f(\mathbf{x})$ with highest total degree, where $\{S_i^\alpha\}$ denotes the collection of all vertex or generalized vertex splines, and $\{\mathbf{u}^j\}$ a set of sample points. When the partition Δ is fairly regular, it is possible to write down a nice formulation of $\{a_{ij}^\alpha\}$. For bivariate C^1 quintic vertex splines, a procedure is outlined in Chui and Lai [64], and for bivariate C^1 quadratic generalized vertex splines, the following formulation is given in Chui and He [60]. Let $\Delta_{MN}^{(1)}$ be a (not necessarily uniform) type-1 triangulation of $\Omega = [a,b] \times [c,d]$ obtained by drawing in all diagonals with positive slopes to the subrectangles $[x_i, x_{i+1}] \times [y_j, y_{j+1}]$ where $a = x_0 < \cdots < x_{M+1} = b$ and $c = y_0 < \cdots < y_{N+1} = d$. Set $f_{ij} = f(x_i, y_j)$. We will use the C^1 quadratic generalized vertex splines S_{ij}^*, T_{ij}^*, and U_{ij}^* at the vertex (x_i, y_j) as given in Example 6.2 in §6.3 that satisfy the δ_n conditions (i) - (iii) in Theorem 6.2 in the following theorem.

THEOREM 6.4. *Let*

$$(6.6) \quad (Qf)(x,y) = \sum_{i=1}^{M}\sum_{j=1}^{N}\left\{f_{ij}S_{ij}^*(x,y) + \left[\frac{f_{ij}-f_{i-1j}}{x_i-x_{i-1}}\right.\right.$$
$$\left.-\frac{f_{i+1j}-f_{i-1j}}{x_{i+1}-x_{i-1}} + \frac{f_{i+1j}-f_{ij}}{x_{i+1}-x_i}\right]T_{ij}^*(x,y)$$
$$\left.+\left[\frac{f_{ij}-f_{ij-1}}{y_j-y_{j-1}} - \frac{f_{ij+1}-f_{ij-1}}{y_{j+1}-y_{j-1}} + \frac{f_{ij+1}-f_{ij}}{y_{j+1}-y_j}\right]U_{ij}^*(x,y)\right\}.$$

Then $(Qf)(x,y) = f(x,y)$ *for all* $f \in \pi_2^2$ *and* (x,y) *in* $[x_1, x_M] \times [y_1, y_N]$.

By a more careful analysis, this result can be extended to the entire region Ω. In addition, an analogous result for an arbitrary triangulation also holds. For more details, see Chui and He [60]. We now turn to least-squares approximation. In the following, Δ will be either a simplicial or parallelepiped partition of a region Ω in $\mathbf{R}^s$. Let $\{V_1, \cdots, V_M\}$ be a vertex spline basis of $\widehat{S}_d^r(\Delta)$ or a generalized vertex spline basis of $S_d^r(\widehat{\Delta})$ when $\widehat{\Delta} \supset \Delta$. To simplify notation, we will denote both $\widehat{S}_d^r(\Delta)$ and $S_d^r(\widehat{\Delta})$ by S_d^r. For $\widehat{S}_d^r(\Delta)$, we will always assume $d \geq 2^s r + 1$, and for $S_d^r(\widehat{\Delta})$, we will only consider the bivariate setting with $r = 1$ and $d = 2$. Both L^2 and ℓ^2 will be considered.

For $f \in L^2(\Omega)$, let $s_f \in S_d^r$ be the best $L^2(\Omega)$ approximant of f from S_d^r and set

$$s_f(\mathbf{x}) = \sum_{i=1}^{M} c_i V_i(\mathbf{x}),$$

$\mathbf{c} = [c_1 \cdots c_M]^T$, and $\mathbf{b} = [b_1 \cdots b_M]^T$, where

$$b_i = \int_\Omega f(\mathbf{x})V_i(\mathbf{x})d\mathbf{x}.$$

Then it follows that the matrix $A = [a_{ij}]$ where

$$a_{ij} = \int_\Omega V_i(\mathbf{x})V_j(\mathbf{x})d\mathbf{x}$$

is nonsingular and

$$\mathbf{c} = A^{-1}\mathbf{b}.$$

The following result is in Chui and Lai [66].

THEOREM 6.5. *Suppose that f is in $C^{d+1}(\Omega)$ if $\triangle$ is a simplicial partition, and is in $C^{sd}(\Omega)$ if $\triangle$ is a parallelepiped partition. Then*

(6.7) $$\| f - s_f \|_{L^2(\Omega)} \leq Ch^{d+1}$$

where C depends only on f and

(6.8) $$h = \max_j (\text{diam } \Omega_j),$$

Ω_j being the cells (i.e., simplices or parallelepipeds) of $\triangle$.

Approximation of discrete data is more complicated but far more important. For data given on a rectangular grid, it is perhaps best to use tensor-product B-splines and the results of Schultz [183] can be applied. In Chui and Lai [66], ℓ^2 approximation of scattered data is considered. Let

$$\{(\mathbf{y}^i, f_i, w_i): \ i = 1, \cdots, L\},$$

where $\mathbf{y}^i \in \Omega$ and $w_i > 0$, be given. The weights $\{w_i\}$ may be normalized so that $\sum_{i=1}^L w_i = 1$. In practice, the quantity of data values f_i at the sample points $\mathbf{y}^i$ is usually very large. In any case, we have $L \geq M$, where M is the dimension of S_d^r. Denote by $\| \ \|_2$ the usual ℓ^2 norm and $\| \ \|_{2,w}$ the weighted ℓ^2 norm with weight $w = \{w_i\}, i = 1, \cdots, L$. If g is a continuous function on Ω, we simply write

$$\| g \|_{2,w} = \| \{g(\mathbf{y}^i)\} \|_{2,w}.$$

In addition, we write $\mathbf{f} = \{f_i\}$.

It is clear that the problem of finding an $s_\mathbf{f} \in S_d^r$, where

(6.9) $$s_\mathbf{f}(x) = \sum_{i=1}^M c_i V_i(\mathbf{x})$$

that satisfies

(6.10) $$\| \mathbf{f} - s_f \|_{2,w} = \inf_{s \in S_d^r} \| \mathbf{f} - s \|_{2,w}$$

does not have a unique solution.

Following Hayes [128], we are interested in solving for

$$\hat{s}_\mathbf{f}(\mathbf{x}) = \sum_{i=1}^M \hat{c}_i V_i(\mathbf{x})$$

that satisfies both the ℓ^2 minimization criterion (6.10) and

$$\| \{\hat{c}_i\} \|_2 = \inf\{\| \{c_i\} \|_2 : \{c_i\} \text{ defines } s_{\mathbf{f}} \text{ as in } (6.9)$$
$$\text{satisfying } (6.10)\}.$$

THEOREM 6.6. *The above problem has a unique solution* $\hat{s}_{\mathbf{f}} \in S_d^r$.

To state the result on the order of approximation in Chui and Lai [66], we need the following notation. It must be emphasized, however, that the order of approximation is only meaningful on a subset E of Ω where the sample points $\mathbf{y}^i$ are fairly *dense*. Hence, we use the notation

$$d_E = \max_{\mathbf{x} \in E} \min_{1 \le i \le L} |\mathbf{x} - \mathbf{y}^i|.$$

In addition, set

$$\delta_E = \min\{w_i \colon \mathbf{y}^i \in E\},$$

and let $\triangle_E$ be the minimum of the radii of the balls inscribed in the (simplicial or parallelepiped) cells that have nonempty intersection with E. We also need the multivariate Markov constant $C(d)$, defined by

$$C(d) = \max_{\substack{\|P_d\|_D = 1 \\ i=1,\cdots,s}} \| \frac{\partial}{\partial x_i} P_d \|_D$$

where D is either the standard simplex

$$\langle 0, \mathbf{e}^1, \cdots, \mathbf{e}^s \rangle$$

with $\mathbf{e}^i = (0, \cdots, 0, 1, 0, \cdots, 0)$ or the unit cube $[0,1]^s$, and the maximum is taken over all polynomials P_d of degree d with unit norm on D. Here d is the total degree if D is the standard simplex, or the coordinate degree if D is the unit cube in $\mathbf{R}^s$. We have the following result.

THEOREM 6.7. *Let E be a subset of Ω with*

(6.11) $$C(d) d_E < \triangle_E.$$

Then for any $f \in C^{d+1}(\Omega)$ *if* $\triangle$ *is a simplicial partition, or* $f \in C^{sd}(\Omega)$ *if* $\triangle$ *is a parallelepiped partition, and* $\mathbf{f} = \{f(\mathbf{y}^i)\}, i = 1, \cdots, L,$

$$\| \mathbf{f} - \hat{s}_{\mathbf{f}} \|_E \le K \left(1 - \frac{C(d)}{\triangle_E} d_E \right)^{-1} (\delta_E)^{-1/2} h^{d+1}$$

where the constant K depends on f.

We remark that under the condition $C(d)d_E < \triangle_E$, the data points must be fairly dense in E.

CHAPTER 7

Computational Algorithms

As we mentioned in Chapter 1, algorithms for computing spline functions may be classified into three types. For multivariate splines $s(\mathbf{x})$ that satisfy some recurrence relationship, such as the box splines as given by (2.6) in Theorem 2.5, the user can compute $s(\mathbf{x})$ *exactly* for each fixed $\mathbf{x}$ using the recurrence relationship and compute $s(\mathbf{y})$ again for another value of $\mathbf{y}$. The second type is to give an efficient *approximation scheme* which is based on another recurrence relationship. This type is very useful for graphic display purposes. The third type is to give an *explicit representation*, say in terms of the Bézier net, of each polynomial piece. Algorithms of this type depend on yet another form of recurrence relationship. In this chapter, we will not go into the first type of algorithm, but discuss briefly a representative algorithm of each of the latter two types. We first consider a simple technique for displaying a polynomial "surface."

7.1. Polynomial surface display. Let $S = \langle \mathbf{x}^0, \cdots, \mathbf{x}^s \rangle$ be an s-simplex (in $\mathbf{R}^s$) and

$$P_n(\mathbf{x}) = \sum_{|\beta|=n} a_\beta \phi_\beta^n(\lambda),$$

where $a_\beta = a_\beta^n$ and $\lambda = (\lambda_0(\mathbf{x}), \cdots, \lambda_s(\mathbf{x}))$, be an nth degree polynomial in terms of the barycentric coordinates λ with respect to this simplex S (cf. §5.1). We are interested in displaying the graph of $P_n(\mathbf{x})$ *efficiently* by giving a good approximation of $P_n(\mathbf{x})$ at a "fairly dense" set of points, which are uniformly spaced on S. The idea is that if the simplex S is very "small" to the human eye, then its Bézier net could be used for its graph. Hence, by subdividing the simplex S into small pieces of subsimplices and giving the Bézier net of the (same) polynomial on each piece, we arrive at a good approximation of the graph of the polynomial.

In computer aided geometric design (CAGD), this is usually accomplished by using de Casteljau's algorithm (cf. de Boor [22], Dahmen [84], and Prautzsch [165]). However, since division by two is attained by a simple shift in binary arithmetic, we suggest "dyadic" subdivision as shown in Figure 7.1 for the two-dimensional setting. For the corner subsimplices, we could use the following relationship, which may be derived from Taylor's formula in Bézier polynomials given in Theorem 6.3.

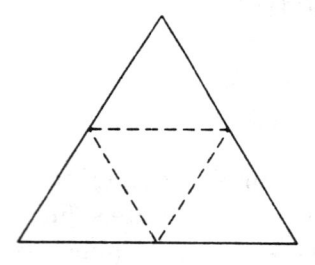

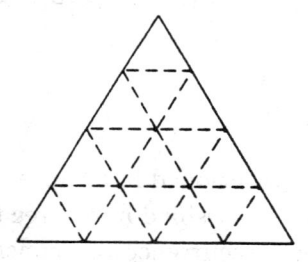

FIG. 7.1

THEOREM 7.1. *Let* $S = \langle \mathbf{x}^0, \cdots, \mathbf{x}^s \rangle$ *be an s-simplex and* $S_1 = \langle \mathbf{x}^0, \frac{1}{2}(\mathbf{x}^0 + \mathbf{x}^1), \cdots, \frac{1}{2}(x^0 + x^s) \rangle$. *Suppose that*

$$P_n(\mathbf{x}) = \sum_{|\alpha|=n} a_\alpha \phi_\alpha^n(\lambda) = \sum_{|\beta|=n} b_\beta \phi_\beta^n(\mu)$$

where $\lambda = (\lambda_0(\mathbf{x}), \cdots, \lambda_s(\mathbf{x}))$ *and* $\mu = (\mu_0(\mathbf{x}), \cdots, \mu_s(\mathbf{x}))$ *are the barycentric coordinates of* $\mathbf{x}$ *relative to the simplices* S *and* S_1, *respectively. Then*

$$(7.1) \qquad b_\beta = \frac{1}{2^{n-\beta_0}} \sum_{\substack{\alpha_i \leq \beta_i, i=1,\cdots,s \\ |\alpha|=n}} \binom{\beta_1}{\alpha_1} \cdots \binom{\beta_s}{\alpha_s} a_\alpha$$

where $\beta = (\beta_0, \cdots, \beta_s) \in \mathbf{Z}_+^{s+1}$ *with* $|\beta| = n$.

To prove this relationship between the Bézier net $\{a_\alpha\}$ on the original simplex S and the Bézier net $\{b_\beta\}$ of the same polynomial on the subsimplex S_1 with a common vertex $\mathbf{x}^0$, we first note that, by Lemma 5.1,

$$D_{s0}^{\gamma_s} \cdots D_{10}^{\gamma_1} P_n(\mathbf{x}) = \frac{n!}{\gamma_0!} \sum_{|\alpha|=\gamma_0} \Delta_{s0}^{\gamma_s} \cdots \Delta_{10}^{\gamma_1} a_\alpha \phi_\alpha^{\gamma_0}(\lambda).$$

Hence, evaluation at $\mathbf{x}^0$ yields

$$D_{s0}^{\gamma_s} \cdots D_{10}^{\gamma_1} P_n(\mathbf{x}^0) = \frac{n!}{\gamma_0!} \Delta_{s0}^{\gamma_s} \cdots \Delta_{10}^{\gamma_1} a_{(\gamma_0,0,\cdots,0)},$$

and it follows that

$$D_{(\mathbf{x}^s+\mathbf{x}^0)/2-\mathbf{x}^0}^{\gamma_s} \cdots D_{(\mathbf{x}^1+\mathbf{x}^0)/2-\mathbf{x}^0}^{\gamma_1} P_n(\mathbf{x}^0)$$
$$= \frac{n!}{\gamma_0!} \frac{1}{2^{\gamma_1+\cdots+\gamma_s}} \Delta_{s0}^{\gamma_s} \cdots \Delta_{10}^{\gamma_1} a_{(\gamma_0,0,\cdots,0)}.$$

COMPUTATIONAL ALGORITHMS

By substituting this data information into Taylor's formula in Theorem 6.3, we have

$$(7.2) \quad b_\beta = \sum_{\substack{\alpha_i \leq \beta_i \\ i=1,\cdots,s}} 2^{-(\alpha_1+\cdots+\alpha_s)} \binom{\beta_1}{\alpha_1} \cdots \binom{\beta_s}{\alpha_s} \\ \times \Delta_{s0}^{\alpha_s} \cdots \Delta_{10}^{\alpha_1} a_{(\alpha_0,0,\cdots,0)},$$

and we must simplify this expression. This can be done one at a time as follows. First, note that

$$\sum_{\alpha_1=0}^{\beta_1} 2^{\beta_1-\alpha_1} \binom{\beta_1}{\alpha_1} \Delta_{10}^{\alpha_1} a_{(\alpha_0,0,\cdots,0)}$$

$$= \sum_{\alpha_1=0}^{\beta_1} \sum_{\gamma_1=0}^{\alpha_1} (-1)^{\gamma_1} 2^{\beta_1-\alpha_1} \binom{\beta_1}{\alpha_1} \\ \times \binom{\alpha_1}{\gamma_1} a_{(n-\alpha_1-\cdots-\alpha_s+\gamma_1,\alpha_1-\gamma_1,0,\cdots,0)}$$

$$= \sum_{\gamma_1=0}^{\beta_1} \sum_{\alpha_1=\gamma_1}^{\beta_1} (-1)^{\gamma_1} 2^{\beta_1-\alpha_1} \binom{\beta_1}{\gamma_1} \\ \times \binom{\beta_1-\gamma_1}{\alpha_1-\gamma_1} a_{(n-\alpha_2-\cdots-\alpha_s-(\alpha_1-\gamma_1),\alpha_1-\gamma_1,0,\cdots,0)}$$

$$= \sum_{\gamma_1=0}^{\beta_1} \sum_{\alpha_1=0}^{\beta_1-\gamma_1} (-1)^{\gamma_1} 2^{\beta_1-\gamma_1-\alpha_1} \binom{\beta_1}{\gamma_1} \\ \times \binom{\beta_1-\gamma_1}{\alpha_1} a_{(n-\alpha_1-\cdots-\alpha_s,\alpha_1,0,\cdots,0)}$$

$$= \sum_{\alpha_1=0}^{\beta_1} \sum_{\gamma_1=0}^{\beta_1-\alpha_1} (-1)^{\gamma_1} 2^{\beta_1-\alpha_1-\gamma_1} \binom{\beta_1}{\alpha_1} \\ \times \binom{\beta_1-\alpha_1}{\gamma_1} a_{(n-\alpha_1-\cdots-\alpha_s,\alpha_1,0,\cdots,0)}$$

$$= \sum_{\alpha_1=0}^{\beta_1} \binom{\beta_1}{\alpha_1} a_{(n-\alpha_1-\cdots-\alpha_s,\alpha_1,0,\cdots,0)}.$$

Hence, substituting this into (7.2) yields

$$b_\beta = \frac{1}{2^{\beta_1}} \sum_{\alpha_1=0}^{\beta_1} \binom{\beta_1}{\alpha_1} \left[\sum_{\substack{\alpha_i \le \beta_i \\ i=2,\cdots,s}} 2^{-(\alpha_2+\cdots+\alpha_s)} \binom{\beta_2}{\alpha_2} \cdots \binom{\beta_s}{\alpha_s} \Delta_{s0}^{\alpha_s} \cdots \Delta_{20}^{\alpha_2} a_{(\alpha_0,\alpha_1,0,\cdots,0)} \right]$$

where the term inside the brackets is the same as the right-hand side of (7.2) with one less summation. Hence, we may conclude by induction that

$$b_\beta = \frac{1}{2^{\beta_1+\cdots+\beta_s}} \sum_{\alpha_1=0}^{\beta_1} \cdots \sum_{\alpha_s=0}^{\beta_s} \binom{\beta_1}{\alpha_1} \cdots \binom{\beta_s}{\alpha_s} a_{(\alpha_0,\cdots,\alpha_s)}$$

$$= \frac{1}{2^{n-\beta_0}} \sum_{\substack{\alpha_i \le \beta_i \\ i=1,\cdots,s}} \binom{\beta_1}{\alpha_1} \cdots \binom{\beta_s}{\alpha_s} a_\alpha.$$

This completes the proof of the theorem.

Example 7.1. Let $n = 3$ in $\mathbf{R}^2$. Then by applying the formula (7.1), we have the following.

$$b_{300} = a_{300}$$
$$b_{210} = \frac{1}{2}(a_{300} + a_{210})$$
$$b_{201} = \frac{1}{2}(a_{300} + a_{201})$$
$$b_{120} = \frac{1}{4}(a_{300} + 2a_{210} + a_{120})$$
$$b_{111} = \frac{1}{4}(a_{300} + a_{210} + a_{201} + a_{111})$$
$$b_{102} = \frac{1}{4}(a_{300} + 2a_{201} + a_{102})$$
$$b_{030} = \frac{1}{8}(a_{300} + 3a_{210} + 3a_{120} + a_{030})$$
$$b_{021} = \frac{1}{8}(a_{300} + 2a_{210} + a_{201} + a_{120} + 2a_{111} + a_{021})$$
$$b_{012} = \frac{1}{8}(a_{300} + a_{210} + 2a_{201} + 2a_{111} + a_{102} + a_{012})$$
$$b_{003} = \frac{1}{8}(a_{300} + 3a_{201} + 3a_{102} + a_{003}).$$

See Figure 7.2.

Of course, the same procedure gives the Bézier nets of the same polynomial relative to the other two corner triangles. To obtain the Bézier net relative to the center subtriangle, we may write it in terms of the Bézier net $\{b_\beta\}$ using the C^3 smoothing condition discussed in §5.3. Since the two triangles form a parallelogram, the formulas are particularly simple (cf. (5.9)). In this example, using $\{c_\gamma\}$ to denote the Bézier net of the same polynomial relative to the center subtriangle (cf. Figure 7.2), we have the following relationship:

$$c_{300} = b_{030}, \quad c_{210} = b_{021}, \quad c_{120} = b_{012}, \quad c_{030} = b_{003}$$
$$c_{201} = (b_{030} + b_{021}) - b_{120},$$
$$c_{111} = (b_{021} + b_{012}) - b_{111}$$
$$c_{021} = (b_{012} + b_{003}) - b_{102}$$
$$c_{102} = (b_{030} + 2b_{021} + b_{012}) - 2(b_{120} + b_{111}) + b_{210}$$
$$c_{012} = (b_{021} + 2b_{012} + b_{003}) - 2(b_{111} + b_{102}) + b_{201}$$
$$c_{003} = (b_{030} + 3b_{021} + 3b_{012} + b_{003}) - 3(b_{120} + 2b_{111} + b_{102})$$
$$\qquad + 3(b_{210} + b_{201}) - b_{300}$$

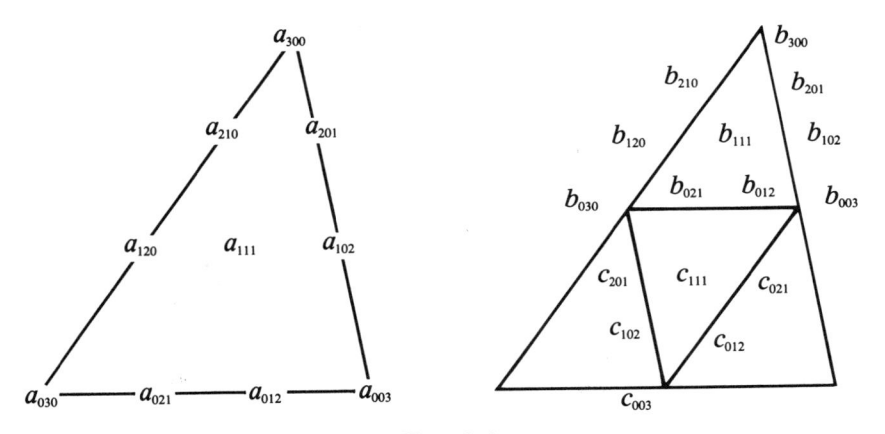

FIG. 7.2

The procedure demonstrated in the above example for taking care of the noncorner subsimplices can be extended to the general setting using the "Pascal triangles." Of course, further subdivision of each subsimplex using the same method will produce a better approximation.

7.2. Discrete box splines. Subdivision algorithms for displaying spline curves discussed in §1.5 can be generalized to display box spline surfaces. These algorithms are developed in Böhm [16], Cohen, Lyche, and Riesenfeld [79], Dahmen and Micchelli [92], [96], and Prautzsch [165]. Following the format of §1.5, we will discuss the line average algorithm (cf. [79] and [96]) in the next section. Here, we will consider the notion of discrete box splines. The box spline $M(\cdot|X_n)$ with direction set

$$X_n = \{\mathbf{x}^1, \cdots, \mathbf{x}^n\} \subset \mathbf{Z}^s \backslash \{\mathbf{0}\}$$

where $\langle X_n \rangle = \mathbf{R}^s$, which was introduced in §2.1, is symmetric with respect to the origin. Hence, without loss of generality, we will assume that the first nonzero component of each $\mathbf{x}^i$ is positive (cf. §2.3). As in the univariate case, however, to describe this algorithm it is more convenient to make a shift, yielding

(7.3) $$B(\mathbf{x}|X_n) = M(\mathbf{x} - \frac{1}{2}\sum_{i=1}^{n}\mathbf{x}^i|X_n)$$

so that the "starting point" is at the origin. In Figure 7.3, we show the supports of some of the typical box splines on the three- and four-directional meshes in $\mathbf{R}^2$:

$$B_{tuvw}(x,y) = M_{tuvw}((x,y) - \frac{1}{2}[t\mathbf{e}^1 + u\mathbf{e}^2 + v(\mathbf{e}^1 + \mathbf{e}^2) + w(\mathbf{e}^2 - \mathbf{e}^1)]).$$

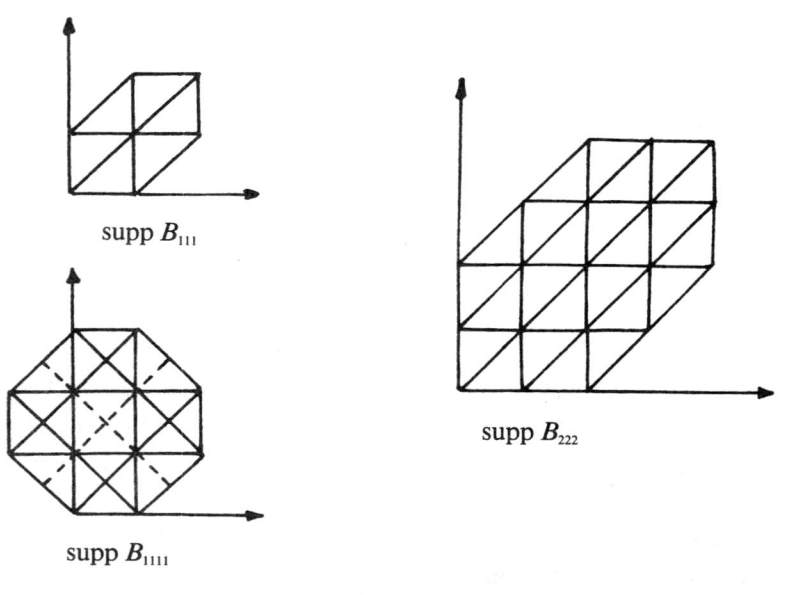

supp B_{111}

supp B_{222}

supp B_{1111}

FIG. 7.3

Recall that in the univariate case, the B-spline $N_n(x)$ is written as a linear combination of the B-splines $N_n(px - j)$ with coefficients $a_p^n(\frac{j}{p})$ on the finer mesh $\frac{1}{p}\mathbf{Z}$, so that the generating function of these coefficients is obtained (cf. (1.15) and (1.16)). Of course, the set of coefficients $\{a_p^n(\frac{j}{p})\}$ provides an approximation of the B-spline $N_n(x)$, and, consequently, its graph. Hence, it may be called a *discrete B-spline*. In the multivariate case, it is a little more involved, since, as we have seen, translates of a single box spline usually cannot generate other locally supported spline functions in the space. That this formulation is still valid can be found in [79] and [96].

We need the following notation. For $\mathbf{y} \in \mathbf{Z}^s \setminus \{\mathbf{0}\}$, we consider the backward difference:

(7.4) $$\nabla_{\mathbf{y}} f(\cdot) = f(\cdot) - f(\cdot - \mathbf{y})$$

and let

$$\nabla_X = \prod_{\mathbf{y} \in X} \nabla_{\mathbf{y}}.$$

The reader might have observed that instead of the central difference used in §2.2, the backward difference is used here because of the shift in (7.3) in defining $B(\mathbf{x}|X_n)$. Hence, another formulation of Theorem 2.9 that expresses the box spline $B(\mathbf{x}|X_n)$ in terms of the multivariate truncated powers $T(\mathbf{x}|X_n)$ may be stated as:

(7.5) $$B(\mathbf{x}|X_n) = \nabla_{X_n} T(\mathbf{x}|X_n).$$

Note that $T(\mathbf{x}|X_n)$ is defined for all $\mathbf{x}$ since the first nonzero component of each $\mathbf{x}^i$ is positive. Now, if the multivariate truncated power is replaced by the discrete truncated power $t(\mathbf{k}|X_n)$ defined in §2.3, the identity (7.5) suggests the following definition of "*discrete box spline.*"

DEFINITION. Let p be a positive integer. The discrete box spline with direction set X_n is defined by

(7.6) $$a_p(\frac{1}{p}\mathbf{k}|X_n) = \frac{1}{p^{n-s}} \nabla_{X_n} t(\frac{1}{p}\mathbf{k}|\frac{1}{p}X_n),$$

where $\mathbf{k} \in \mathbf{Z}^s$.

Recall the notation of the discrete affine cone defined in §2.3 by

$$c(X_n) = \left\{ \sum_{i=1}^n t_i \mathbf{x}^i : \ t_i \in Z_+, \ i = 1, \cdots, n \right\}.$$

100   CHAPTER 7

The following result can be found in [79] and [96].

THEOREM 7.2.

(7.7) $$B(\mathbf{x}|X_n) = \sum_{\mathbf{k}\in c(X_n)} a_p(\frac{1}{p}\mathbf{k}|X_n)B(p\mathbf{x} - \mathbf{k}|X_n).$$

Hence, indeed, a box spline can be generated by translates of its refinement. To prove (7.7), we recall from Theorem 2.8 that

$$\int_{\mathbf{R}^s} T(\mathbf{x}|X_n)f(\mathbf{x})d\mathbf{x} = \int_{[0,\infty)^n} f(\sum_{i=1}^n t_i\mathbf{x}^i)dt_1\cdots dt_n,$$

for any continuous function $f(\mathbf{x})$ with compact support. Hence, we have

$$\int_{\mathbf{R}^s} \frac{1}{p^n}T(\mathbf{x}|\frac{1}{p}X_n)f(\mathbf{x})d\mathbf{x}$$

$$= \frac{1}{p^n}\int_{[0,\infty)^n} f(\sum_{i=1}^n t_i\frac{1}{p}\mathbf{x}^i)dt_1\cdots dt_n$$

$$= \frac{1}{p^n}\int_{[0,\infty)^n} f(\sum_{i=1}^n \tau_i\mathbf{x}^i)(pd\tau_1)\cdots(pd\tau_n)$$

$$= \int_{[0,\infty)^n} f(\sum_{i=1}^n \tau_i\mathbf{x}^i)d\tau_1\cdots d\tau_n$$

so that

$$\int_{\mathbf{R}^s} [T(\mathbf{x}|X_n) - \frac{1}{p^n}T(\mathbf{x}|\frac{1}{p}X_n)]f(\mathbf{x})d\mathbf{x} = 0$$

for any test function $f(\mathbf{x})$. This gives

(7.8) $$\frac{1}{p^n}T(\mathbf{x}|\frac{1}{p}X_n) = T(\mathbf{x}|X_n).$$

On the other hand, a reformulation of (2.11) gives:

$$T(\mathbf{x}|X_n) = \sum_{\mathbf{k}\in c(X_n)} t(\mathbf{k}|X_n)B(\mathbf{x} - \mathbf{k}|X_n).$$

Combining this with (7.8), we have:

$$T(\mathbf{x}|X_n) = \frac{1}{p^n}T(\mathbf{x}|\frac{1}{p}X_n)$$

$$= \frac{1}{p^n}\sum_{\mathbf{z}\in c(\frac{1}{p}X_n)} t(\mathbf{z}|\frac{1}{p}X_n)B(\mathbf{x} - \mathbf{z}|\frac{1}{p}X_n)$$

$$= \frac{1}{p^n}\sum_{\mathbf{k}\in c(X_n)} t(\frac{1}{p}\mathbf{k}|\frac{1}{p}X_n)B(\mathbf{x} - \frac{1}{p}\mathbf{k}|\frac{1}{p}X_n).$$

Hence, by (7.5) and summation by parts, we have:

(7.9)
$$B(\mathbf{x}|X_n) = \nabla_{\mathbf{x}_n} T(\mathbf{x}|X_n)$$
$$= \frac{1}{p^n} \sum_{\mathbf{k} \in c(X_n)} t(\frac{1}{p}\mathbf{k}|\frac{1}{p}X_n) \nabla_{\mathbf{x}_n} B(\mathbf{x} - \frac{1}{p}\mathbf{k}|\frac{1}{p}X_n)$$
$$= \frac{1}{p^n} \sum_{\mathbf{k} \in c(X_n)} \nabla_{\mathbf{x}_n} t(\frac{1}{p}\mathbf{k}|\frac{1}{p}X_n) B(\mathbf{x} - \frac{1}{p}\mathbf{k}|\frac{1}{p}X_n).$$

Now, let us apply Theorem 2.7 with

$$A = \frac{1}{p} I_s = \begin{bmatrix} \frac{1}{p} & & 0 \\ & \ddots & \\ 0 & & \frac{1}{p} \end{bmatrix}$$

to yield

$$B(\mathbf{x} - \frac{1}{p}\mathbf{k}|\frac{1}{p}X_n)$$
$$= B(\mathbf{x} - \frac{1}{p}\mathbf{k}|AX_n)$$
$$= B(A(p\mathbf{x} - \mathbf{k})|AX_n)$$
$$= \frac{1}{\det A} B(p\mathbf{x} - \mathbf{k}|X_n)$$
$$= p^s B(p\mathbf{x} - \mathbf{k}|X_n).$$

Therefore, (7.9) may be written as

$$B(\mathbf{x}|X_n) = \frac{1}{p^{n-s}} \sum_{\mathbf{k} \in c(X_n)} \nabla_{\mathbf{x}_n} t(\frac{1}{p}\mathbf{k}|\frac{1}{p}X_n) B(p\mathbf{x} - \mathbf{k}|X_n)$$

and this is (7.7) by using the definition of the discrete box spline in (7.6). This completes the proof of Theorem 7.2.

We now take the Fourier transform of both sides of (7.7), using the fact that

$$\int_{\mathbf{R}^s} f(p\mathbf{y} - \mathbf{k}) e^{-i\mathbf{y}\cdot\mathbf{x}} d\mathbf{y} = \frac{1}{p^s} e^{-i\frac{1}{p}\mathbf{k}\cdot\mathbf{x}} \hat{f}(\frac{1}{p}\mathbf{x}),$$

to obtain:

(7.10) $$\hat{B}(\mathbf{x}|X_n) = \sum_{\mathbf{k} \in c(X_n)} a_p(\frac{1}{p}\mathbf{k}|X_n) \frac{1}{p^s} e^{-i\frac{1}{p}\mathbf{k}\cdot\mathbf{x}} \hat{B}(\frac{1}{p}\mathbf{x}|X_n).$$

On the other hand, we also have, from (2.3),

$$\frac{\widehat{B}(\mathbf{x}|X_n)}{\widehat{B}(\frac{1}{p}\mathbf{x}|X_n)} = \prod_{j=1}^{n} \left(\frac{1-e^{-i\mathbf{x}\cdot\mathbf{x}^j}}{i\mathbf{x}\cdot\mathbf{x}^j}\right) \left(\frac{1-e^{-i\frac{1}{p}\mathbf{x}\cdot\mathbf{x}^j}}{i\frac{1}{p}\mathbf{x}\cdot\mathbf{x}^j}\right)^{-1}.$$

Hence, by setting $\mathbf{x} = (x_1, \cdots, x_s)$ and

$$\mathbf{z} = (e^{-i\frac{1}{p}x_1}, \cdots, e^{-i\frac{1}{p}x_s}),$$

(7.10) can be reformulated to give the following result.

THEOREM 7.3.

(7.11) $$\frac{1}{p^{n-s}} \prod_{j=1}^{n} \frac{1-\mathbf{z}^{p\mathbf{x}^j}}{1-\mathbf{z}^{\mathbf{x}^j}} = \sum_{\mathbf{k} \in c(X_n)} a_p(\frac{1}{p}\mathbf{k}|X_n)\mathbf{z}^{\mathbf{k}}.$$

The function on the left-hand side of the identity (7.11) is called the generating function of the discrete box spline. It will be discussed in the next section that, analogous to the univariate case, the discrete box spline $\{a_p(\frac{1}{p}\mathbf{k}|X_n)\}$ approximates the box spline $B(\mathbf{x}|X_n)$ at $\mathbf{x} = \frac{1}{p}\mathbf{k}$, and hence, provides the graph of $B(\mathbf{x}|X_n)$ when p is chosen to be a very large positive integer.

7.3. The line average algorithm. Throughout, we will always assume that the direction set X_n contains the standard basis

$$X_s = \{\mathbf{e}^1, \cdots, \mathbf{e}^s\}$$

of $\mathbf{R}^s$. That is, $\mathbf{x}^i = \mathbf{e}^i$ for $i = 1, \cdots, s$. It follows from (7.11) that since

$$\frac{1}{p^{s-s}} \prod_{j=1}^{s} \frac{1-\mathbf{z}^{p\mathbf{e}^j}}{1-\mathbf{z}^{\mathbf{e}^j}}$$

$$= \prod_{j=1}^{s} \frac{1-z_j^p}{1-z_j}$$

$$= \prod_{j=1}^{s} (1 + \cdots + z_j^{p-1})$$

$$= \sum_{0 \leq k_i \leq p-1} z_1^{k_1} \cdots z_s^{k_s},$$

we have:

(7.12) $$a_p(\frac{1}{p}\mathbf{k}|X_s) = \begin{cases} 1 & \text{for } 0 \leq k_i \leq p-1 \\ 0 & \text{otherwise.} \end{cases}$$

To obtain $a_p(\frac{1}{p}\mathbf{k}|X_n)$ from (7.12), we simply take the average:

(7.13) $$a_p(\frac{1}{p}\mathbf{k}|\mathbf{x}^1,\cdots,\mathbf{x}^{m+1}) = \frac{1}{p}\sum_{j=0}^{p-1} a_p(\frac{1}{p}(\mathbf{k}-j\mathbf{x}^{m+1})|\mathbf{x}^1,\cdots,\mathbf{x}^m)$$

for $m = s, \cdots, n-1$. The formulas (7.12) and (7.13) together may be called the *line average algorithm* for computing the discrete box splines that approximate the box spline $B(\mathbf{x}|X_n)$ at $\mathbf{x} = \frac{1}{p}\mathbf{k}$ and provides its graph. To arrive at (7.13), we simply note that, by mathematical induction,

$$\sum_{\mathbf{j}\in c(X_{m+1})} a_p(\frac{1}{p}\mathbf{j}|\mathbf{x}^1,\cdots,\mathbf{x}^{m+1})\mathbf{z}^{\mathbf{j}}$$

$$= \frac{1}{p^{m+1-s}} \prod_{j=1}^{m+1} \frac{1-\mathbf{z}^{p\mathbf{x}^j}}{1-\mathbf{z}^{\mathbf{x}^j}} = \frac{1}{p^{m-s}} \prod_{j=1}^{m} \frac{1-\mathbf{z}^{p\mathbf{x}^j}}{1-\mathbf{z}^{\mathbf{x}^j}} \left(\frac{1}{p}\sum_{\ell=0}^{p-1} \mathbf{z}^{\ell\mathbf{x}^{m+1}}\right)$$

$$= \sum_{\mathbf{k}\in c(X_m)} a_p(\frac{1}{p}\mathbf{k}|X_m) \left(\frac{1}{p}\sum_{\ell=0}^{p-1} \mathbf{z}^{\mathbf{k}+\ell\mathbf{x}^{m+1}}\right)$$

$$= \sum_{\mathbf{j}\in c(X_{m+1})} \left\{\frac{1}{p}\sum_{\ell=0}^{p-1} a_p(\frac{1}{p}(\mathbf{j}-\ell\mathbf{x}^{m+1})|X_m)\right\} \mathbf{z}^{\mathbf{j}}.$$

Let us now consider an arbitrary box spline series

(7.14) $$s(\mathbf{x}|X_n) = \sum_{\mathbf{j}\in \mathbf{Z}^s} c_{\mathbf{j}} B(\mathbf{x}-\mathbf{j}|X_n).$$

By Theorem 7.2, we have

$$s(\mathbf{x}|X_n) = \sum_{\mathbf{j}\in \mathbf{Z}^s} c_{\mathbf{j}} \left\{\sum_{\mathbf{k}\in c(X_n)} a_p(\frac{1}{p}\mathbf{k}|X_n) B(p(\mathbf{x}-\mathbf{j})-\mathbf{k}|X_n)\right\}$$

$$= \sum_{\mathbf{j}\in \mathbf{Z}^s} c_{\mathbf{j}} \sum_{\mathbf{m}\in c(X_n)} a_p(\frac{1}{p}(\mathbf{m}-p\mathbf{j})|X_n) B(p\mathbf{x}-\mathbf{m}|X_n)$$

$$= \sum_{\mathbf{m}\in c(X_n)} \left\{\sum_{\mathbf{j}\in \mathbf{Z}^s} c_{\mathbf{j}} a_p(\frac{1}{p}\mathbf{m}-\mathbf{j}|X_n)\right\} B(p\mathbf{x}-\mathbf{m}|X_n).$$

Hence, using (7.12) and (7.13) we arrive at the following theorem.

THEOREM 7.4. *Let*

$$\text{(7.15)} \qquad d_p(\tfrac{1}{p}\mathbf{k}|X_n) = \sum_{\mathbf{j}\in\mathbf{Z}^s} a_p(\tfrac{1}{p}\mathbf{k}-\mathbf{j}|X_n)c_{\mathbf{j}}.$$

Then the box spline series (7.14) has the representation

$$\text{(7.16)} \qquad s(\mathbf{x}) = \sum_{\mathbf{k}\in c(X_n)} d_p(\tfrac{1}{p}\mathbf{k}|X_n)B(p\mathbf{x}-\mathbf{k}|X_n).$$

To compute (7.15), start with

$$d_p(\mathbf{m}+\tfrac{1}{p}\mathbf{k}|X_s) = c_{\mathbf{m}}$$

for $0 \le k_i \le p-1, i=1,\cdots,s$ and $\mathbf{m}\in\mathbf{Z}^s$. Then for $u=s,\cdots,n-1$, compute

$$d_p(\tfrac{1}{p}\mathbf{k}|X_{u+1}) = \frac{1}{p}\sum_{\ell=0}^{p-1} d_p(\tfrac{1}{p}(\mathbf{k}-\ell\mathbf{x}^{u+1})|X_u).$$

Of course the "discrete spline" $\{d_p(\tfrac{1}{p}\mathbf{k}|X_n)\}$ approximates the box spline series $s(\mathbf{x})$ at $\mathbf{x}=\tfrac{1}{p}\mathbf{k}$ for all $\mathbf{k}\in\mathbf{Z}^s$ and large values of $p\in\mathbf{Z}_+$ and the convergence rate is $O(p^{-1})$ (cf. Dahmen and Micchelli [92]). However, if the distinct vectors in the direction set X_n are linearly independent, then analogous to the univariate result mentioned in Chapter 1, the convergence can be shown to be quadratic, provided an appropriate shift is made, but cannot be improved to $o(p^{-2})$ in general (cf. Dahmen [85] and Dahmen, Dyn, and Levin [86]).

Suppose that an approximation has already been made, and we would like to improve the approximation. For this purpose, the following formula can be used. For positive integers p and q,

$$\text{(7.17)} \qquad d_{pq}(\tfrac{1}{pq}\mathbf{k}|X_n) = \sum_{\mathbf{j}\in c(X_n)} a_q(\tfrac{1}{q}\mathbf{k}-\mathbf{j}|X_n)d_p(\tfrac{1}{p}\mathbf{j}|X_n).$$

In particular, to "double the resolution" of the graph, we may use the algorithm:

$$\text{(7.18)} \qquad \begin{cases} a_2(\tfrac{1}{2}\mathbf{k}|X_n) = \dfrac{1}{2^{n-s}} \displaystyle\sum_{\substack{\mathbf{x}^{j_1}+\cdots+\mathbf{x}^{j_i}=\mathbf{k} \\ 1\le j_1,\cdots,j_i \le n \\ 1\le i \le n}} 1 \\ d_{2p}(\tfrac{1}{2p}\mathbf{k}|X_n) = \displaystyle\sum_{\mathbf{j}\in c(X_n)} a_2(\tfrac{1}{2}\mathbf{k}-\mathbf{j}|X_n)d_p(\tfrac{1}{p}\mathbf{j}|X_n). \end{cases}$$

7.4. Bézier nets of locally supported splines. For analytical purposes, it is sometimes necessary to know the exact formulation of the polynomial pieces of a multivariate locally supported spline. Algorithms for this purpose have been classified under the third type. In this section, we will present an algorithm for computing the Bézier nets of box splines and minimally supported splines on the three-directional mesh in $\mathbf{R}^2$. This algorithm can be generalized to the other situations but the formulation would have to be more involved. The reader is referred to Chui and Lai [64] where box splines are computed and Chui and He [57] for computing minimally supported splines. The algorithm to be presented here will be valid for both situations, at least on the three-directional mesh.

As in §3.3, we let A and B be the triangular regions with vertices $(0,0), (1,1), (0,1)$ and $(0,0), (1,0), (1,1)$, respectively. Also set

(7.19) $\quad g_{1,00}^{111} = B_{111}, \quad g_{2,00}^{011} = \chi_A, \quad g_{3,00}^{101} = \chi_B.$

(Note that in (3.10) they are denoted by g_1^0, g_2^0, g_3^0, respectively, but here we need the superscript and extra subscript notation to describe our computational scheme.) For better understanding, the reader is referred to §1.5 where a computational scheme of the Bernstein nets of B-splines is discussed.

Let us first recall the bivariate convolution operator notation $I_1 f$, $I_2 f$, and $I_3 f$ introduced in §3.3. Now, for each k, $k = 1, 2, 3$, define

(7.20) $\quad \begin{cases} g_{k,ij}^{m+1nu} = I_1 g_{k,ij}^{mnu} \\ g_{k,ij}^{mn+1u} = I_2 g_{k,ij}^{mnu} \\ g_{k,ij}^{mnu+1} = I_3 g_{k,ij}^{mnu} \end{cases}$

where

$$g_{k,ij}^{mnu}(x,y) = g_{k,00}^{mnu}(x-i, y-j).$$

It is clear that each $g_{k,ij}^{mnu}$ is a piecewise polynomial of total degree $(m+n+u-2)$ on the three-directional mesh $\Delta^{(1)}$, and that

$$g_{1,00}^{mnu} = B_{mnu} = B_{mnu0}$$

are the "shifted" box splines defined in (3.5). We also recall that

$\begin{cases} g_1^k = g_{1,00}^{k+1,k+1,k+1} & \in S_{3k+1}^{2k}(\Delta^{(1)}) \\ g_2^k = g_{2,00}^{k,k+1,k+1} & \in S_{3k}^{2k-1}(\Delta^{(1)}) \\ g_3^k = g_{3,00}^{k+1,k,k+1} & \in S_{3k}^{2k-1}(\Delta^{(1)}) \end{cases}$

are all the minimally supported splines in the corresponding spaces (see (3.14)). The algorithm to be described in this section is designed to compute all the $g_{k,ij}^{mnu}$. We need the following lemma.

LEMMA 7.1. *for each* k, $k = 1, 2, 3,$

(7.21)
$$\begin{cases} \dfrac{\partial}{\partial x} g_{k,ij}^{m+1nu}(x,y) = g_{k,ij}^{mnu}(x,y) - g_{k,ij}^{mnu}(x-1,y) \\ \qquad\qquad\qquad = g_{k,ij}^{mnu}(x,y) - g_{k,i+1j}^{mnu}(x,y) \\ \dfrac{\partial}{\partial y} g_{k,ij}^{mn+1u}(x,y) = g_{k,ij}^{mnu}(x,y) - g_{k,ij}^{mnu}(x,y-1) \\ \qquad\qquad\qquad = g_{k,ij}^{mnu}(x,y) - g_{k,ij+1}^{mnu}(x,y) \\ \dfrac{\partial}{\partial x} g_{k,ij}^{mnu+1}(x,y) = g_{k,ij}^{mnu}(x,y) - g_{k,ij}^{mnu}(x-1,y-1) \\ \qquad\qquad\qquad = g_{k,ij}^{mnu}(x,y) - g_{k,i+1\ j+1}^{mnu}(x,y). \end{cases}$$

This lemma follows from the definition (7.20); namely,

$$\begin{cases} g_{k,ij}^{m+1nu}(x,y) = \int_{x-1}^{x} g_{k,ij}^{mnu}(t,y)dt \\ g_{k,ij}^{mn+1u}(x,y) = \int_{y-1}^{y} g_{k,ij}^{mnu}(x,t)dt \\ g_{k,ij}^{mnu+1}(x,y) = \int_{x-1}^{x} g_{k,ij}^{mnu}(t,t+y-x)dt. \end{cases}$$

In Figure 7.4, the vertices of the triangles are $(i+p, j+q), (i+p, j+q+1), (i+p+1, j+q+1)$ and $(i+p, j+q), (i+p+1, j+q), (i+p+1, j+q+1)$, respectively, where the ordering for the first triangle (i.e., the upper one) is in the clockwise direction and that of the second triangle (i.e., the lower one) is in the counterclockwise direction and both start from the same point $(i+p, j+q)$. These orientations reflect our notation of the Bézier

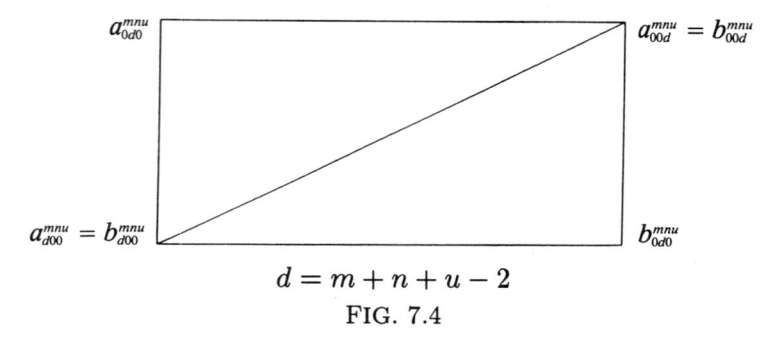

a_{0d0}^{mnu} \qquad\qquad $a_{00d}^{mnu} = b_{00d}^{mnu}$

$a_{d00}^{mnu} = b_{d00}^{mnu}$ \qquad\qquad b_{0d0}^{mnu}

$d = m + n + u - 2$

FIG. 7.4

nets of $g_{k,ij}^{mnu}$. Hence, by denoting these Bézier nets on these two triangles with $\{a_{rst}^{mnu}(i,j,p,q)\}$ and $\{b_{rst}^{mnu}(i,j,p,q)\}$, respectively, we can now decide their positions. We have chosen these orientations in order to preserve the sign of the directional derivatives (see Figure 7.4). Here, $d = m + n + u - 2$ denotes the degree of the polynomial pieces. Note that we have omitted the subscript k only to simplify notation. The following initial values for m, n, and u, however, do depend on k as in Figures 7.5–7.7 below:

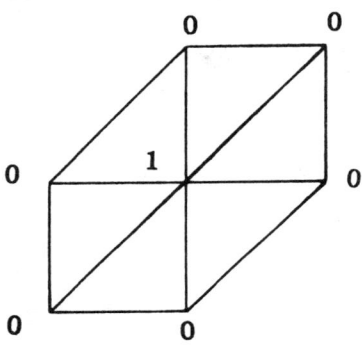

(i) $k = 1, m = n = u = 1$

FIG. 7.5

(ii) $k = 2, m = 0, n = u = 1$

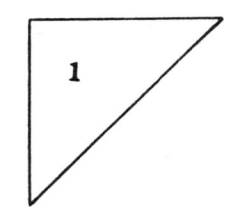

FIG. 7.6

(iii) $k = 3, m = 1, n = 0, u = 1$

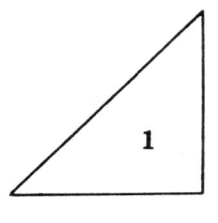

FIG. 7.7

With the above initial Bézier nets, we can now compute any $g_{k,ij}^{mnu}$ using the following algorithm, which is a simple consequence of Lemma 7.1.

THEOREM 7.5. *Set*

$$a_{rst}^{mnu}(i,j,-1,q) = a_{rst}^{mnu}(i,j,p,-1)$$
$$= b_{rst}^{mnu}(i,j,-1,q) = b_{rst}^{mnu}(i,j,p,-1) = 0.$$

I. *For* $t = 0, \cdots, d$ *and* $r+s+t = d$,

$$a_{r\ s\ t+1}^{m+1\ n\ u}(i,j,p,q) = a_{r\ s+1\ t}^{m+1\ n\ u}(i,j,p,q)$$
$$+ \frac{1}{m+n+u-1}[a_{rst}^{mnu}(i,j,p,q) - a_{rst}^{mnu}(i+1,j,p-1,q)]$$

where $a_{r\ s+1\ 0}^{m+1\ n\ u}(i,j,p,q) = b_{0\ r\ s+1}^{m+1nu}(i,j,p-1,q)$.

II. *For* $s = 0, \cdots, d$ *and* $r+s+t = d$,

$$b_{r\ s+1\ t}^{m+1\ n\ u}(i,j,p,q) = b_{r+1\ s\ t}^{m+1\ n\ u}(i,j,p,q)$$
$$+ \frac{1}{m+n+u-1}[b_{rst}^{mnu}(i,j,p,q) - b_{rst}^{mnu}(i+1,j,p-1,q)]$$

where $b_{r+1\ 0\ t}^{m+1\ n\ u}(i,j,p,q) = a_{r+1\ 0\ t}^{m+1\ n\ u}(i,j,p,q)$.

III. *For* $s = 0, \cdots, d$ *and* $r+s+t = d$,

$$a_{r\ s+1\ t}^{m\ n+1\ u}(i,j,p,q) = a_{r+1\ s\ t}^{m\ n+1\ u}(i,j,p,q)$$
$$+ \frac{1}{m+n+u-1}[a_{rst}^{mnu}(i,j,p,q) - a_{rst}^{mnu}(i,j+1,p,q-1)]$$

where $a_{r+1\ 0\ t}^{m\ n+1\ u}(i,j,p,q) = b_{r+1\ 0\ t}^{m\ n+1\ u}(i,j,p,q)$.

IV. *For* $t = 0, \cdots, d$ *and* $r+s+t = d$,

$$b_{r\ s\ t+1}^{m\ n+1\ u}(i,j,p,q) = b_{r\ s+1\ t}^{m\ n+1\ u}(i,j,p,q)$$
$$+ \frac{1}{m+n+u-1}[b_{rst}^{mnu}(i,j,p,q) - b_{rst}^{mnu}(i,j+1,p,q-1)]$$

where $b_{r\ s+1\ 0}^{m\ n+1\ u}(i,j,p,q) = a_{0\ r\ s+1}^{m\ n+1\ u}(i,j,p,q-1)$.

V. *For* $s = 0, \cdots, d$ *and* $r+s+t = d$,

$$a_{r\ s\ t+1}^{m\ n\ u+1}(i,j,p,q) = a_{r+1\ s\ t}^{m\ n\ u+1}(i,j,p,q)$$
$$+ \frac{1}{m+n+u-1}[a_{rst}^{mnu}(i,j,p,q) - a_{rst}^{mnu}(i+1,j+1,p-1,q-1)]$$

where $a_{r\ s+1\ 0}^{m\ n\ u+1}(i,j,p,q) = b_{0\ r\ s+1}^{mnu+1}(i,j,p-1,q)$.

VI. *For* $s = 0, \cdots, d$ *and* $r+s+t = d$,

$$b^{m\ n\ u+1}_{r\ s\ t+1}(i,j,p,q) = b^{m\ n\ u+1}_{r+1\ s\ t}(i,j,p,q)$$

$$+\frac{1}{m+n+u-1}[b^{mnu}_{rst}(i,j,p,q) - b^{mnu}_{rst}(i+1,j+1,p-1,q-1)]$$

where $b^{m\ n\ u+1}_{r\ s+1\ 0}(i,j,p,q) = a^{m\ n\ u+1}_{0\ r\ s+1}(i,j,p,q-1)$.

In Figures 7.8–7.12, we demonstrate this procedure by computing the Bézier nets of the box splines $B_{211} = g^{211}_{1,00}, B_{221} = g^{221}_{1,00}, B_{112} = g^{112}_{1,00}, B_{122} = g^{122}_{1,00}$, and $B_{222} = g^{222}_{1,00}$. As we mentioned earlier, this procedure can be applied to compute box splines as well as minimally and quasi-minimally supported splines on the four-directional mesh (see Chui and Lai [64] for more examples).

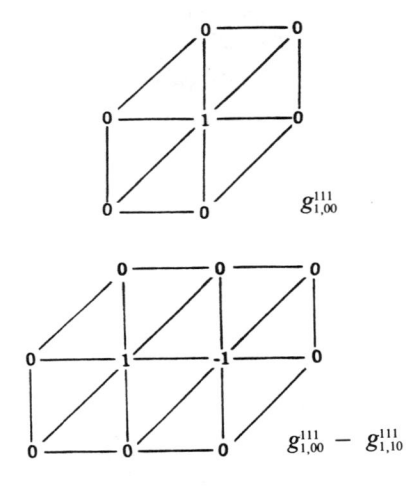

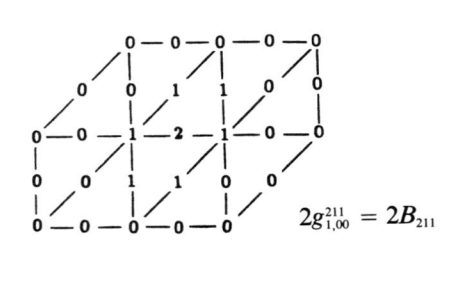

FIG. 7.8

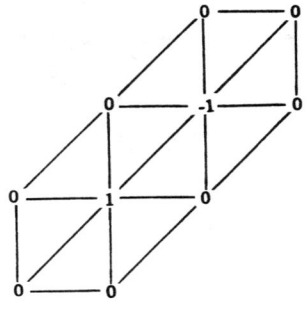

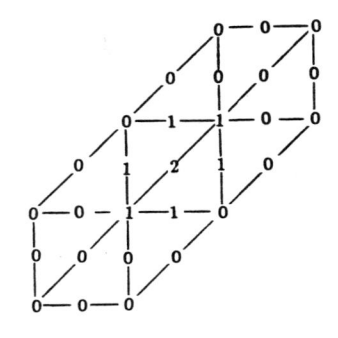

FIG. 7.9

110   CHAPTER 7

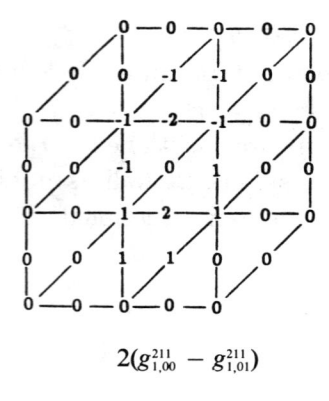

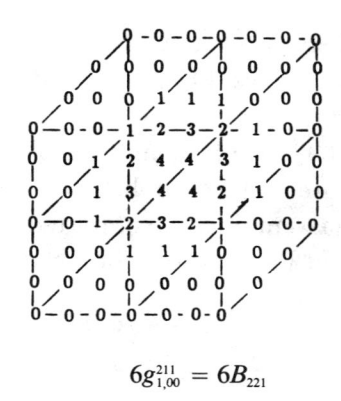

$2(g^{211}_{1,00} - g^{211}_{1,01})$ $6g^{211}_{1,00} = 6B_{221}$

FIG. 7.10

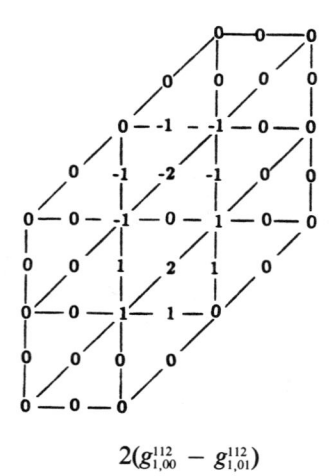

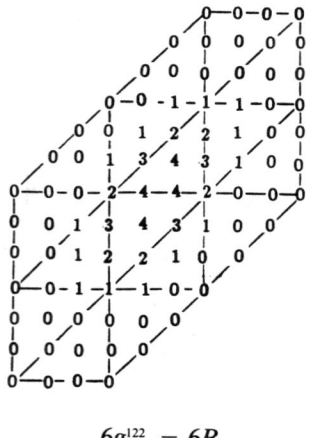

$2(g^{112}_{1,00} - g^{112}_{1,01})$ $6g^{122}_{1,00} = 6B_{122}$

FIG. 7.11

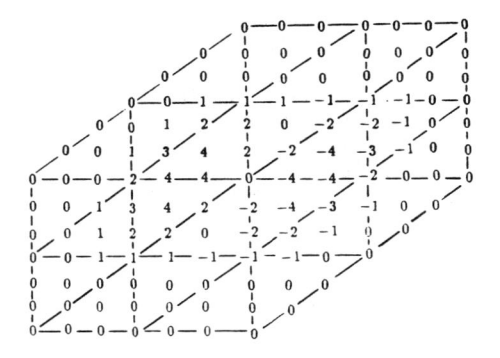

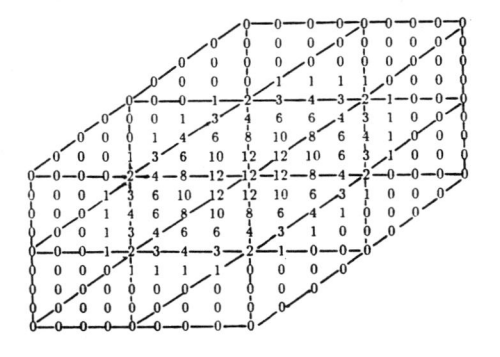

FIG. 7.12

CHAPTER 8

Quasi-Interpolation Schemes

The problem of constructing an approximation formula in the form of a multivariate spline series that guarantees the highest possible order of approximation is the focal point of our discussion in this chapter. As we mentioned earlier (cf. §1.4 and §6.4), the central idea is to construct a spline series that reproduces polynomials up to the highest possible degree. For this reason, a multivariate analogue of the univariate polynomial reproducing identity due to Marsden [153] as stated in Theorem 1.5 would be a very useful tool for our study. For multivariate splines in $S_d^r(\Delta)$ where Δ is an arbitrary partition, this problem is exceedingly difficult, since constructing locally supported spline functions is already a very difficult task. There seems to be only two feasible situations in general: one is to allow the degree d to be large, say $d \geq 2^s r + 1$, in which case multivariate vertex splines discussed in §6.1 provide an useful tool, and another is to allow certain symmetric refinement of Δ so that lower degree generalized vertex splines introduced in §6.2 can be constructed. Both approaches are intimately related to the methods of finite elements, and hence, have many important applications.

This chapter is devoted to the study of approximation of discrete data taken on an uniform grid, so that spline series in terms of translates of certain appropriate locally supported functions can be used for constructing the quasi-interpolants. In other words, locally supported functions $\phi_1, \cdots, \phi_k$ are used, and linear functionals $\lambda_{\mathbf{i}}^{(1)}, \cdots, \lambda_{\mathbf{i}}^{(k)}$ are constructed so that an approximation scheme of the type

$$(8.1) \quad \sum_{\mathbf{i}} \lambda_{\mathbf{i}}^{(1)}(\sigma_{h^{-1}}f)\phi_1(\frac{1}{h}\mathbf{x}-\mathbf{i}) + \cdots + \sum_{\mathbf{i}} \lambda_{\mathbf{i}}^{(k)}(\sigma_{h^{-1}}f)\phi_k(\frac{1}{h}\mathbf{x}-\mathbf{i})$$

can be obtained. Unfortunately, there doesn't seem to be any available method for constructing this type of quasi-interpolation formulas, except for the special case $k = 1$. Hence, we will restrict our attention only to spline series of the form

$$(8.2) \qquad Q(f) = \sum_{\mathbf{i}} \lambda_{\mathbf{i}}(f)\phi(\mathbf{x}-\mathbf{i}).$$

Here, there are two degrees of freedom in this formulation: one being the possibility of choosing the "best" locally supported function ϕ in the given multivariate spline space, and the other being the freedom of supplying the linear functionals $\{\lambda_{\mathbf{i}}\}$ that should give an efficient and yet stable and robust approximation scheme.

We will characterize the "optimality" of ϕ by the so-called Strang and Fix conditions and will introduce the notion of the *commutator* of ϕ whose order can be shown to be equivalent to that of the Strang and Fix conditions via the Poisson summation formula. Two methods to construct the sequence of linear functionals $\lambda_{\mathbf{i}}$ will be discussed.

8.1. The commutator operator. We first recall the following so-called Poisson summation formula.

THEOREM 8.1. *Let ϕ be a continuous function in $\mathbf{R}^s$ having compact support, such that its Fourier transform satisfies $\{\hat{\phi}(2\pi\mathbf{j})\} \in \ell^1(\mathbf{Z}^s)$. Then*

$$(8.3) \qquad \sum_{\mathbf{j}\in\mathbf{Z}^s} \phi(\mathbf{j}) = \sum_{\mathbf{j}\in\mathbf{Z}^s} \hat{\phi}(2\pi\mathbf{j}).$$

As an application, let us apply (8.3) to the functions

$$\mathbf{t}^\alpha \phi(\mathbf{x}-\mathbf{t}) \quad \text{and} \quad (\mathbf{x}-\mathbf{t})^\alpha \phi(\mathbf{t})$$

where $\mathbf{x} \in \mathbf{R}^s$ is fixed and $\alpha \in \mathbf{Z}^s_+$, and obtain the following.

COROLLARY 8.1. *Let ϕ be a compactly supported continuous function in $\mathbf{R}^s$ with $\{\hat{\phi}(2\pi\mathbf{j})\} \in \ell^1(\mathbf{Z}^s)$. Then for any $\alpha \in \mathbf{Z}^s_+$,*

$$(8.4) \qquad \sum_{\mathbf{j}\in\mathbf{Z}^s} \mathbf{j}^\alpha \phi(\mathbf{x}-\mathbf{j}) = \sum_{\mathbf{j}\in\mathbf{Z}^s} (-iD)^\alpha (e^{i\mathbf{x}\cdot\mathbf{y}} \hat{\phi}(\mathbf{y}))\big|_{\mathbf{y}=2\pi\mathbf{j}}$$

and

$$(8.5) \qquad \sum_{\mathbf{j}\in\mathbf{Z}^s} (\mathbf{x}-\mathbf{j})^\alpha \phi(\mathbf{j}) = \sum_{\mathbf{j}\in\mathbf{Z}^s} e^{-i2\pi\mathbf{j}\cdot\mathbf{x}} (-iD)^\alpha (e^{i\mathbf{x}\cdot\mathbf{y}} \hat{\phi}(\mathbf{y}))\big|_{\mathbf{y}=2\pi\mathbf{j}}$$

for all $\mathbf{x} \in \mathbf{R}^s$. Here, the derivatives are taken with respect to $\mathbf{y}$.

It is obvious that if $\mathbf{x} \in \mathbf{Z}^s$, then (8.4) and (8.5) agree, being simply the discrete convolution (or convolution over the group $\mathbf{Z}^s$) of $\{\phi(\mathbf{j})\}$ with $\{\mathbf{j}^\alpha\}$. However, it is also clear from the right-hand sides of (8.4) and (8.5) that they are usually different. This leads to the notion of the commutator of ϕ introduced in Chui, Jetter, and Ward [63].

DEFINITION. The commutator of a compactly supported continuous function ϕ in $\mathbf{R}^s$ is an operator on $C(\mathbf{R}^s)$ defined by

$$(8.6) \quad [\phi|f](\mathbf{x}) = \sum_{\mathbf{j} \in \mathbf{Z}^s} \phi(\mathbf{x} - \mathbf{j}) f(\mathbf{j}) - \sum_{\mathbf{j} \in \mathbf{Z}^s} f(\mathbf{x} - \mathbf{j}) \phi(\mathbf{j}).$$

Note that if $f \in \pi^s$, the space of all polynomials in s variables, then $[\phi|f]$ vanishing identically would mean that

$$\sum_{\mathbf{j} \in \mathbf{Z}^s} f(\mathbf{x} - \mathbf{j}) \phi(\mathbf{j}) = \sum_{\mathbf{j} \in \mathbf{Z}^s} f(\mathbf{j}) \phi(\mathbf{x} - \mathbf{j})$$

where the left-hand side is a polynomial and the right-hand side is a linear combination of the translates of ϕ over $\mathbf{Z}^s$.

Throughout, we will use the following notation for the monomials:

$$(8.7) \quad m_\alpha(\mathbf{x}) = \frac{1}{\alpha!} \mathbf{x}^\alpha, \quad \alpha \in \mathbf{Z}_+^s.$$

We have the following result.

THEOREM 8.2. *Let ϕ be a compactly supported continuous function in $\mathbf{R}^s$ with $\{\hat{\phi}(2\pi \mathbf{j})\} \in \ell^1(\mathbf{Z}^s)$ and $\alpha \in \mathbf{Z}_+^s$. Then*

$$(8.8) \quad [\phi|m_\alpha](\mathbf{x}) = \sum_{\mathbf{j} \in \mathbf{Z}^s} \left\{ \sum_{0 \leq \beta \leq \alpha} (-i)^{|\beta|} m_{\alpha-\beta}(\mathbf{x}) \frac{1}{\beta!} D^\beta \hat{\phi}(2\pi \mathbf{j}) \right\} (e^{i 2\pi \mathbf{j} \cdot \mathbf{x}} - 1)$$

for all $\mathbf{x} \in \mathbf{R}^s$. Furthermore,

$$(8.9) \quad [\phi|m_\gamma](\mathbf{x}) = 0, \quad \text{all } \mathbf{x} \in \mathbf{R}^s, \; \gamma \leq \alpha$$

if and only if

$$(8.10) \quad D^\gamma \hat{\phi}(2\pi \mathbf{j}) = 0, \quad \text{all } \mathbf{j} \in \mathbf{Z}^s \setminus \{\mathbf{0}\}, \; \gamma \leq \alpha.$$

The identity (8.8) is an immediate consequence of Corollary 8.1. That (8.10) implies (8.9) is trivial. To show that (8.9) implies (8.10), we may use mathematical induction on α. Indeed, since

$$(8.11) \qquad [\phi|m_0](\mathbf{x}) = \sum_{0 \neq \mathbf{j} \in \mathbf{Z}^s} \hat{\phi}(2\pi\mathbf{j})e^{i2\pi\mathbf{j}\cdot\mathbf{x}} + \hat{\phi}(\mathbf{0}) - \sum_{\mathbf{j} \in \mathbf{Z}^s} \phi(\mathbf{j})$$

so that $[\phi|m_0] = 0$ implies $\hat{\phi}(2\pi\mathbf{j}) = 0$ for all $\mathbf{j} \in \mathbf{Z}^s \setminus \{\mathbf{0}\}$, the induction procedure can be initiated. Suppose that (8.9) holds. Then by the induction hypothesis, we have

$$D^\gamma \hat{\phi}(2\pi\mathbf{j}) = 0, \quad \mathbf{j} \in \mathbf{Z}^s \setminus \{\mathbf{0}\},$$

for all γ, with $0 \leq \gamma \leq \alpha$ and $\gamma \neq \alpha$, so that (8.8) gives

$$(8.12) \qquad \sum_{0 \neq \mathbf{j} \in \mathbf{Z}^s} D^\alpha \hat{\phi}(2\pi\mathbf{j})e^{i2\pi\mathbf{j}\cdot\mathbf{x}} - \sum_{0 \neq \mathbf{j} \in \mathbf{Z}^s} D^\alpha \hat{\phi}(2\pi\mathbf{j}) = 0,$$

or $D^\alpha \hat{\phi}(2\pi\mathbf{j}) = 0$, $\mathbf{j} \in \mathbf{Z}^s \setminus \{\mathbf{0}\}$. This completes the proof of the theorem.

We now introduce the notion of the degree of the commutator of ϕ (cf. Chui, Jetter, and Ward [63]) and the so-called S-F conditions (cf. Fix and Strang [109] and Dahmen and Micchelli [89]).

DEFINITION. The commutator of a compactly supported continuous function ϕ in $\mathbf{R}^s$ with $\{\hat{\phi}(2\pi\mathbf{j})\} \in \ell^1(\mathbf{Z}^s)$ is said to have *commutator index* $\alpha \in \mathbf{Z}_+^s$ if (8.9) holds. The collection of all commutator indices α of ϕ is called the *commutator indicator set* of ϕ, denoted by Γ_ϕ. (Note that Γ_ϕ is a lower set, by which we mean that if $\beta \in \Gamma$ and $\mathbf{0} \leq \alpha \leq \beta$ then $\alpha \in \Gamma$.) The largest integer n for which $\alpha \in \Gamma_\phi$ whenever $|\alpha| \leq n$ is called the *degree of the commutator* of ϕ.

DEFINITION. A compactly supported continuous function ϕ in $\mathbf{R}^s$ with $\{\hat{\phi}(2\pi\mathbf{j})\} \in \ell^1(\mathbf{Z}^s)$ is said to satisfy the S-F condition with S-F index $\alpha \in \mathbf{Z}_+^s$ if

$$(8.13) \qquad \begin{cases} \hat{\phi}(\mathbf{0}) = 1 \quad \text{and} \\ D^\gamma \hat{\phi}(2\pi\mathbf{j}) = 0, \quad \mathbf{j} \in \mathbf{Z}^s \setminus \{\mathbf{0}\}, \ \gamma \leq \alpha. \end{cases}$$

The collection of all S-F indices of ϕ is called the S-F *indicator set* of ϕ, denoted by Λ_ϕ. The largest n for which $\alpha \in \Lambda_\phi$ whenever $|\alpha| \leq n$ is called the S-F degree of ϕ.

As usual, in both of the above definitions, the *order* is one larger than the degree.

Suppose that $[\phi|1] \equiv 0$, or equivalently, $\hat{\phi}(2\pi\mathbf{j}) = 0$ for all $\mathbf{j} \in \mathbf{Z}^s\backslash\{\mathbf{0}\}$ by (8.11). Then the condition

$$\hat{\phi}(\mathbf{0}) = 1 \tag{8.14}$$

in (8.13) is equivalent to the condition

$$\sum_{\mathbf{j} \in \mathbf{Z}^s} \phi(\mathbf{j}) = 1 \tag{8.15}$$

by applying the Poisson summation formula (8.3). We will say that ϕ is *normalized* if (8.15) is satisfied. In addition, since $[\phi|1] \equiv 0$, the normalization condition (8.15) is equivalent to

$$\sum_{\mathbf{j} \in \mathbf{Z}^s} \phi(\mathbf{x} - \mathbf{j}) = 1$$

for all $\mathbf{x} \in \mathbf{R}^s$. Hence, this normalization is important in that it guarantees a *partition of unity* by using all the translates of ϕ over $\mathbf{Z}^s$. As a consequence of Theorem 8.2, we have the following result.

THEOREM 8.3. *Let ϕ be a compactly supported continuous function in $\mathbf{R}^s$ with $\{\hat{\phi}(2\pi\mathbf{j})\} \in \ell^1(\mathbf{Z}^s)$ such that (8.15) is satisfied. Then the commutator indicator set and the S-F indicator set of ϕ are identical; that is, $\Gamma_\phi = \Lambda_\phi$. In particular, the commutator of ϕ has order m if and only if ϕ satisfies the S-F condition of order m.*

For further information and applications of the commutator, the reader is referred to de Boor [23], Chui, Jetter, and Ward [63], Chui and Lai [65], ter Morsche [160], Ward [204], and the survey article by Jetter [138].

8.2. Polynomial-generating formulas. Following de Boor [23], we consider the Appell sequence $\{g_\alpha\}, \alpha \in \mathbf{Z}_+^s$, of a bounded linear functional μ on $C(\mathbf{R}^s)$ that satisfies $\mu(m_0) = 1$, where in general, $m_\alpha(\mathbf{x})$ is the monomial of coordinate degree $\alpha \in \mathbf{Z}_+^s$ defined in (8.7). That is, $\{g_\alpha\}$ is defined by:

$$\begin{cases} g_\alpha \text{ is a polynomial of coordinate degree } \alpha \\ \mu(D^\beta g_\alpha) = \delta_{\beta\alpha} \end{cases} \tag{8.16}$$

where $\delta_{\beta\alpha}$ denotes the Kronecker delta. By writing

$$g_\alpha(\mathbf{x}) = \sum_{\gamma \leq \alpha} a_\gamma m_\gamma(\mathbf{x}),$$

we note that

$$\mu(D^\beta g_\alpha) = \sum_{\gamma \leq \alpha} a_\gamma \mu(m_{\gamma-\beta})$$

where $m_{\gamma-\beta} = 0$ if $\gamma - \beta \notin \mathbf{Z}_+^s$. Hence, the coefficient matrix $[\mu(m_{\gamma-\beta})]$ in the solution of $\{a_\gamma\}$ is triangular and has unit diagonal elements, so that g_α is uniquely determined by (8.16). The following identity can be easily verified.

LEMMA 8.1.

(8.17) $$g_\alpha(\mathbf{x}) = m_\alpha(\mathbf{x}) - \sum_{\substack{\beta \leq \alpha \\ \beta \neq \alpha}} \mu(m_{\alpha-\beta}) g_\beta(\mathbf{x})$$

for all $\mathbf{x} \in \mathbf{R}^s$.

Since $g_0(\mathbf{x}) = m_0(\mathbf{x}) = 1$, the above formula provides an inductive scheme for computing $g_\alpha(\mathbf{x})$.

As in the above section, let ϕ be a compactly supported continuous function in $\mathbf{R}^s$ with $\{\hat{\phi}(2\pi\mathbf{j})\} \in \ell^1(\mathbf{Z}^s)$. Suppose that the bounded linear functional μ on $C(\mathbf{R}^s)$ is defined by

(8.18) $$\mu(f) = \sum_{\mathbf{j} \in \mathbf{Z}^s} f(\mathbf{j}) \phi(-\mathbf{j}).$$

Then for any $\alpha \in \Gamma_\phi$, where Γ_ϕ denotes the commutator indicator set of ϕ, we have

$$\sum_{\mathbf{j} \in \mathbf{Z}^s} g_\alpha(\mathbf{j}) \phi(\mathbf{x}-\mathbf{j}) = \sum_{\mathbf{j} \in \mathbf{Z}^s} g_\alpha(\mathbf{x}-\mathbf{j}) \phi(\mathbf{j}),$$

which is a polynomial of coordinate degree α, so that

$$D^\beta \left[\sum_{\mathbf{j} \in \mathbf{Z}^s} g_\alpha(\mathbf{j}) \phi(\mathbf{x}-\mathbf{j}) \right] \bigg|_{\mathbf{x}=0}$$

$$= \sum_{\mathbf{j} \in \mathbf{Z}^s} D^\beta g_\alpha(-\mathbf{j}) \phi(\mathbf{j})$$

$$= \sum_{\mathbf{j} \in \mathbf{Z}^s} D^\beta g_\alpha(\mathbf{j}) \phi(-\mathbf{j})$$

$$= \mu(D^\beta g_\alpha) = \delta_{\beta\alpha}.$$

Since the monomial $m_\alpha(\mathbf{x})$ is the only polynomial of coordinate degree α that satisfies the property $D^\beta m_\alpha(\mathbf{0}) = \delta_{\beta\alpha}$, we have arrived at the following polynomial-generating formula given in Chui, Jetter, and Ward [63].

THEOREM 8.4. *Let ϕ be a compactly supported continuous function in $\mathbf{R}^s$ with $\{\hat\phi(2\pi\mathbf{j})\} \in \ell^1(\mathbf{Z}^s)$ and commutator indicator set Γ_ϕ. Set $g_0(\mathbf{x}) = 1$ and define $g_\beta(\mathbf{x})$ inductively by*

$$(8.19) \qquad g_\beta(\mathbf{x}) = m_\beta(\mathbf{x}) - \sum_{\mathbf{j} \in \mathbf{Z}^s} \phi(\mathbf{j}) \sum_{\substack{\gamma \leq \beta \\ \gamma \neq \beta}} \frac{(-\mathbf{j})^{\beta-\gamma}}{(\beta-\gamma)!} g_\gamma(\mathbf{x})$$

for $\beta \in \mathbf{Z}^s$. Then

$$(8.20) \qquad m_\alpha(\mathbf{x}) = \sum_{\mathbf{k} \in \mathbf{Z}^s} g_\alpha(\mathbf{k})\phi(\mathbf{x}-\mathbf{k})$$

for all $\mathbf{x} \in \mathbf{R}^s$ and $\alpha \in \Gamma_\phi$.

Note that in computing the polynomials $g_\beta(\mathbf{x})$ in (8.19), it is necessary to know the values of $\phi(\mathbf{j})$, $\mathbf{j} \in \mathbf{Z}^s$. If ϕ is a minimally or quasi-minimally supported spline, a box spline, or any linear combination of them, these values can be determined by using the method introduced in §7.4 or the recurrence relationships (3.16), (3.17), or (2.6). For box splines, however, it is perhaps easier to compute $D^\beta \hat\phi(2\pi\mathbf{j})$, where the Fourier transform $\hat\phi$ is given by a linear combination of the expressions given in (2.3). By applying (8.4), we may reformulate $g_\beta(\mathbf{x})$ in (8.19) as follows:

$$(8.21) \qquad g_\beta(\mathbf{x}) = m_\beta(\mathbf{x}) - \sum_{\substack{\gamma \leq \beta \\ \gamma \neq \beta}} \frac{1}{(\beta-\gamma)!} (-iD)^{\beta-\gamma} \hat\phi(\mathbf{0}) g_\gamma(\mathbf{x}).$$

Hence, if ϕ is chosen to satisfy $D^\gamma \hat\phi(\mathbf{0}) = 0, \mathbf{0} \neq \gamma \leq \beta$, then g_β agrees with m_β and (8.20) yields a very simple polynomial-generating formula. It is perhaps interesting to compare (8.20) using (8.21) with the univariate result in (1.12). The following relationships on $g_\beta(\mathbf{x})$ are also useful.

LEMMA 8.2. *Let $\beta \in \mathbf{Z}_+^s$. Then*

$$(8.22) \qquad D^\gamma g_\beta(\mathbf{x}) = g_{\beta-\gamma}(\mathbf{x})$$

for all $\gamma \leq \beta$, *and*

(8.23) $$g_\beta(\mathbf{x}) = \sum_{\gamma \leq \beta} g_{\beta-\gamma}(\mathbf{0}) m_\gamma(\mathbf{x})$$

for all $\mathbf{x} \in \mathbf{R}^s$.

The identity (8.22) follows easily from the definition of the Appell sequence, and (8.23) is simply the Taylor formula of $g_\beta(\mathbf{x})$ at the origin. For more details along this direction, the reader is referred to Chui and Lai [65], where the following multivariate analogue of Marsden's identity (cf. (1.11)) is also formulated.

THEOREM 8.5. *Let ϕ be a compactly supported continuous function in* $\mathbf{R}^s$ *with* $\{\hat{\phi}(2\pi \mathbf{j})\} \in \ell^1(\mathbf{Z}^s)$ *so that (8.15) is satisfied, and let Γ_ϕ denote its commutator indicator set. Then for any* $\alpha \in \Gamma_\phi$,

(8.24) $$\frac{1}{\alpha!}(\mathbf{x} - \mathbf{y})^\alpha = \sum_{\mathbf{j} \in \mathbf{Z}^s} g_\alpha(\mathbf{j} - \mathbf{y})\phi(\mathbf{x} - \mathbf{j})$$

for all $\mathbf{x}, \mathbf{y} \in \mathbf{R}^s$.

To prove this result, we appeal to (8.22) and consider the following differentiation of order β with respect to $\mathbf{y}$:

(8.25)
$$D^\beta \left\{ m_\alpha(\mathbf{x} - \mathbf{y}) - \sum_{\mathbf{j} \in \mathbf{Z}^s} g_\alpha(\mathbf{j} - \mathbf{y})\phi(\mathbf{x} - \mathbf{j}) \right\}$$
$$= (-1)^{|\beta|} \left\{ m_{\alpha-\beta}(\mathbf{x} - \mathbf{y}) - \sum_{\mathbf{j} \in \mathbf{Z}^s} g_{\alpha-\beta}(\mathbf{j} - \mathbf{y})\phi(\mathbf{x} - \mathbf{j}) \right\}.$$

Now (8.24) follows by mathematical induction. Indeed, by the induction hypothesis, (8.25) implies that the quantity

$$m_\alpha(\mathbf{x} - \mathbf{y}) - \sum_{\mathbf{j} \in \mathbf{Z}^s} g_\alpha(\mathbf{j} - \mathbf{y})\phi(\mathbf{x} - \mathbf{j})$$

is independent of $\mathbf{y}$. Since it is zero at $\mathbf{y} = \mathbf{0}$ by (8.20), we have established (8.24).

To demonstrate the utility of the formulation (8.21) of $g_\beta(\mathbf{x})$, we consider the case where ϕ is a box spline on the three-directional mesh, shifted as in §7.1 so that the "starting point" is at the origin. The Fourier transform of the box spline B_{tuv} is then

$$\widehat{B}_{tuv}(x,y) = p^t(x)p^u(y)p^v(x+y)$$

where

$$p(t) = \frac{1-e^{-it}}{it}.$$

The commutator indicator set of B_{tuv} can be shown to consist of the commutator indices

$$(t+v-w_1, w_1-1)$$

for all w_1 with $1 \leq w_1 \leq \min(u, t+v)$ and

$$(w_2 - 1, u+v-w_2)$$

for all w_2 with $1 \leq w_2 \leq \min(t, u+v)$. To compute $D^\gamma \widehat{B}_{tuv}(0,0)$, we set

$$\begin{cases} A_j^1 = \dfrac{1}{(j+1)!} \\ \text{and} \\ A_j^{q+1} = \dfrac{q+1}{j+q+1} \sum_{\mu=0}^{j} \dfrac{1}{\mu!} A_{j-\mu}^q \end{cases}$$

where $j \in \mathbf{Z}_+$ and $q = 1, 2, \cdots$. Then for $\alpha \in \Gamma_{B_{tuv}}$ and $\gamma = (\gamma_1, \gamma_2) \leq \alpha$, we have

$$D^\gamma \widehat{B}_{tuv}(0,0) = \gamma! \sum_{\mu=0}^{\gamma_1} \sum_{\nu=0}^{\gamma_2} (-i)^{|\gamma|} \frac{(\mu+\nu)!}{\mu!\nu!} A_{\gamma_1-\mu}^t A_{\gamma_2-\nu}^u A_{\mu+\nu}^v.$$

Proof of the above result and further details can be found in Chui and Lai [65].

8.3. Construction of quasi-interpolants. In this section ϕ will always denote a compactly supported continuous function in $\mathbf{R}^s$ which satisfies $\{\hat{\phi}(2\pi \mathbf{j})\} \in \ell^1(\mathbf{Z}^s)$ and the normalization condition (8.15). As in the case of a box spline, we consider the ϕ-series

$$\sum_{\mathbf{j} \in \mathbf{Z}^s} c_{\mathbf{j}} \phi(\cdot - \mathbf{j})$$

and denote this vector space by $S(\phi)$. For any $h > 0$, consider the scaling operator

$$(\sigma_h f)(\mathbf{x}) = f(\frac{1}{h}\mathbf{x})$$

and set

$$S_h(\phi) = \{\sigma_h f : f \in S(\phi)\}.$$

As in §2.4, we say that $S(\phi)$ has approximation order m, if m is the largest integer for which

$$\text{dist}_{L^\infty}(f, S_h(\phi)) = O(h^m)$$

for all sufficiently smooth functions f. In Strang and Fix [197], it is shown that the approximation order of $S(\phi)$ is given by the S-F order (i.e., S-F degree + 1) of ϕ.

Remarks.

(1) In view of the S-F condition (8.13) and the formula of the Fourier transform of a box spline, this result can be used to verify the approximation order result in (2.12).

(2) Suppose that $S_d^r(\Delta)$ is some multivariate spline space in $C^r(\mathbf{R}^s)$ of degree d and with a regular partition Δ given by some direction set X_n, say. To construct the "best" ϕ, we may consider a basis $\mathcal{B}$ of the subspace of locally supported functions in $S_d^r(\Delta)$ and let ϕ be a linear combination of the elements in $\mathcal{B}$, choosing the coefficients that yield the largest S-F degree. By using the minimally and quasi-minimally supported functions or box splines for $\mathcal{B}$, this procedure is feasible since their Fourier transforms have simple formulations.

In this section we will improve the statement of the result of Strang and Fix, and in addition, present a constructive proof which can be found in Chui and Lai [65].

Let Γ be a finite lower set in $\mathbf{Z}_+^s$ and π_Γ^s the space of polynomials of the form

$$p(\mathbf{x}) = \sum_{\beta \in \Gamma} b_\beta \mathbf{x}^\beta.$$

By a result of Hakopian [126], we know that for any given data $\{f_\mathbf{j} : \mathbf{j} \in \Gamma\}$, there is a unique $p \in \pi_\Gamma^s$ such that $p(\mathbf{j}) = f_\mathbf{j}$ for all $\mathbf{j} \in \Gamma$. Hence, the matrix $[\mathbf{j}^\beta]$, where $\mathbf{j}, \beta \in \Gamma$, is nonsingular. (We are quite sure that this must be a well-known result in linear algebra but we are not aware of any reference to it. For the special case where the lower set is given by $|\alpha| \le n$ for some $n \in \mathbf{Z}_+$, the determinant of $[\mathbf{j}^\beta]$ can be evaluated by using (9.6), and in the univariate setting, the matrix is actually totally positive.) This observation guarantees the existence and uniqueness of the sequence $\{a_\mathbf{j}\}, \mathbf{j} \in \Gamma$, that satisfies

$$\sum_{\mathbf{j}\in\Gamma} a_\mathbf{j}\mathbf{j}^\beta = \beta! g_\beta(\mathbf{0}), \quad \beta \in \Gamma \tag{8.26}$$

where $g_\beta(\mathbf{x})$ is a polynomial of coordinate degree β given by (8.19). Another interesting property of $g_\beta(\mathbf{x})$ is the following.

LEMMA 8.3.

$$\sum_{\mathbf{j}\in\Gamma} a_\mathbf{j}(\mathbf{k}+\mathbf{j})^\beta = \beta! g_\beta(\mathbf{k}) \tag{8.27}$$

for all $\mathbf{k} \in \mathbf{Z}^s$ and $\beta \in \Gamma$.

This result is an immediate consequence of (8.26), (8.22), and (8.23). Indeed, for any $\mathbf{k} \in \mathbf{Z}^s$,

$$\sum_{\mathbf{j}\in\Gamma} a_\mathbf{j}(\mathbf{k}+\mathbf{j})^\beta = \sum_{\gamma\leq\beta} \binom{\beta}{\gamma} \mathbf{k}^\gamma \sum_{\mathbf{j}\in\Gamma} a_\mathbf{j}\mathbf{j}^{\beta-\gamma}$$

$$= \sum_{\gamma\leq\beta} \frac{\beta!}{\gamma!} \mathbf{k}^\gamma g_{\beta-\gamma}(\mathbf{0})$$

$$= \beta! g_\beta(\mathbf{k}).$$

The coefficients $\{a_\mathbf{j}\}$ defined in (8.26) yield the sequence of linear functionals $\{\lambda_\mathbf{k}\}$, which we define as follows: Again, let Γ be a lower set, and consider

$$\lambda_\mathbf{k}(g) = \sum_{\mathbf{j}\in\Gamma} a_\mathbf{j} g(\mathbf{k}+\mathbf{j}). \tag{8.28}$$

Of course, if $g = \sigma_{h^{-1}}(f)$, then

$$\lambda_\mathbf{k}(\sigma_{h^{-1}} f) = \sum_{\mathbf{j}\in\Gamma} a_\mathbf{j} f(h(\mathbf{k}+\mathbf{j})), \tag{8.29}$$

and we arrive at the following "quasi-interpolation" operator

$$(Q_\Gamma^h f)(\mathbf{x}) = \sum_{\mathbf{k}\in\mathbf{Z}^s} \lambda_\mathbf{k}(\sigma_{h^{-1}} f)\phi(\frac{1}{h}\mathbf{x}-\mathbf{k}). \tag{8.30}$$

We have the following result.

THEOREM 8.6. *Let ϕ be a compactly supported continuous function in $\mathbf{R}^s$ satisfying $\{\hat{\phi}(2\pi \mathbf{j})\} \in \ell^1(\mathbf{Z}^s)$ and (8.15), and let $\Gamma = \Gamma_\phi$ be its commutator indicator set. Then $Q_\Gamma^h p \equiv p$ for all $p \in \pi_\Gamma^s$.*

To prove this result, we simply consider $\beta \in \Gamma$ and apply (8.27) and (8.20) consecutively, yielding

$$(Q_\Gamma^h m_\beta)(\mathbf{x}) = \sum_{\mathbf{k} \in \mathbf{Z}^s} \frac{1}{\beta!} \left\{ \sum_{\mathbf{j} \in \Gamma} a_{\mathbf{j}}(\mathbf{k} + \mathbf{j})^\beta \right\} h^{|\beta|} \phi(\frac{1}{h}\mathbf{x} - \mathbf{k})$$

$$= h^{|\beta|} \sum_{\mathbf{k} \in \mathbf{Z}^s} g_\beta(\mathbf{k}) \phi(\frac{1}{h}\mathbf{x} - \mathbf{k})$$

$$= h^{|\beta|} m_\beta(\frac{1}{h}\mathbf{x}) = m_\beta(\mathbf{x}).$$

Remark. The linear functionals $\lambda_\mathbf{k}$ defined in (8.28) are simply translates of a single linear functional $\lambda = \lambda_0$ in the sense that

$$\lambda(g(\cdot + \mathbf{k})) = \lambda_\mathbf{k}(g).$$

Since the support of λ is $\Gamma \subset \mathbf{Z}_+^s$, the "quasi-interpolation" operator is not symmetric. To shift the support of λ, by some $\mathbf{k}_0 \in \mathbf{Z}_+^s$, say, it is necessary to define a new set of coefficients $\{\tilde{a}_\mathbf{j}\}$ using the linear system

$$\sum_{\mathbf{j} \in \Gamma} \tilde{a}_\mathbf{j}(\mathbf{j} - \mathbf{k}_0)^\beta = \beta! g_\beta(\mathbf{0})$$

instead of (8.26). It is then clear that

$$\sum_{\mathbf{j} \in \Gamma} \tilde{a}_\mathbf{j}(\mathbf{k} + \mathbf{j} - \mathbf{k}_0)^\beta = \beta! g_\beta(\mathbf{k})$$

is satisfied, so that we may define

$$\tilde{\lambda}_\mathbf{k}(g) = \sum_{\mathbf{j} \in \mathbf{Z}^s} \tilde{a}_\mathbf{j} g(\mathbf{k} + \mathbf{j} - \mathbf{k}_0),$$

which yields a "more symmetric" quasi-interpolation operator

$$(\widetilde{Q}_\Gamma^h f)(\mathbf{x}) = \sum_{\mathbf{k} \in \mathbf{Z}^s} \tilde{\lambda}_\mathbf{k}(\sigma_{h^{-1}} f) \phi(\frac{1}{h}\mathbf{x} - \mathbf{k}).$$

It is now not difficult to show that the linear operators Q_Γ^h and $\widetilde{Q}_\Gamma^h$ provide the optimal order of approximation from $S(\phi)$. In fact, since we have made use of the commutator indicator set $\Gamma = \Gamma_\phi$, we can say a little more.

As usual, let $\{\mathbf{e}^i\}$ be the standard basis of $\mathbf{R}^s$; that is, $\mathbf{e}^i = (0, \cdots, 0, 1, 0, \cdots, 0)$, and consider

$$\widehat{\Gamma} = \bigcup_{\alpha \in \Gamma} \bigcup_{1 \leq i \leq s} \{\beta : \beta \leq \alpha + \mathbf{e}^i\}.$$

Let $C^{\widehat{\Gamma}}$ denote the class of functions f with continuous $D^\gamma f$ for all $\gamma \in \widehat{\Gamma}$. We have the following result.

THEOREM 8.7. *There exists a positive constant K depending only on ϕ such that for all f in $C^{\widehat{\Gamma}}$,*

(8.31) $$\| Q_\Gamma^h f - f \|_F \leq K \sum_{\beta \in \widehat{\Gamma} \setminus \Gamma} \| D^\beta f \|_F \, h^{|\beta|}.$$

In particular,

(8.32) $$\| Q_\Gamma^h f - f \|_F = O(h^m)$$

where m is the approximation order of $S(\phi)$. Here, the supremum norm over any compact subset F of $\mathbf{R}^s$ is used.

Recall that $m = n + 1$, where n is the commutator degree or S-F degree of ϕ. The proof of the above result is standard in approximation theory by using Theorem 8.6 and the Taylor expansion of f. This theorem justifies that Q_Γ^h is indeed a quasi-interpolation operator.

8.4. Neumann series approach. In this section, we will give a sequence of quasi-interpolation formulas introduced in Chui and Diamond [52] by considering partial sums of the formal Neumann series expansion of some "interpolation operator" which will be defined more precisely later. Again, let ϕ be any compactly supported continuous function in $\mathbf{R}^s$ with $\{\hat{\phi}(2\pi \mathbf{j})\} \in \ell^1(\mathbf{Z}^s)$ and with commutator indicator set Γ_ϕ and commutator degree $n \in \mathbf{Z}_+$. In determining our quasi-interpolation operators, however, we must know the values of $\phi(\mathbf{j})$, for all $\mathbf{j} \in \mathbf{Z}_+^s$, and as usual, ϕ is normalized to satisfy (8.15).

For a given data sequence $F = \{f(\mathbf{j})\}, \mathbf{j} \in \mathbf{Z}^s$, if we wish to determine an $s_f \in S(\phi)$ that interpolates the data on $\mathbf{Z}^s$, we would write

$$s_f(\mathbf{x}) = \sum_{\mathbf{k} \in \mathbf{Z}^s} c_{\mathbf{k}} \phi(\mathbf{x} - \mathbf{k})$$

and consider the linear system $s_f(\mathbf{j}) = f(\mathbf{j})$, or

(8.33) $$\sum_{\mathbf{k} \in \mathbf{Z}^s} c_{\mathbf{k}} \phi(\mathbf{j} - \mathbf{k}) = f(\mathbf{j}), \quad \mathbf{j} \in \mathbf{Z}^s.$$

By setting $C = \{c_{\mathbf{k}}\}$, $\Phi = \{\phi(\mathbf{j})\}$, and using the notation $C\Phi$ to represent convolution of C with Φ on $\mathbf{Z}^s$, (8.33) may be written as

$$(I - M)C = F$$

where $M = I - \Phi$ and $I = \{\delta_{\mathbf{j0}}\}$ is the convolution identity. Hence, the "interpolation operator" C has the formal expression

$$C = (I - M)^{-1}F$$

and the formal convolution inverse $(I - M)^{-1}$ has the formal Neumann series expansion

(8.34) $$(I - M)^{-1} = I + M + M^2 + \cdots.$$

Although the above two expressions are purely symbolic, the partial sums of the series in (8.34), denoted by

$$\wedge_k = (I + M + \cdots + M^k),$$

are certainly well defined. This enables us to define the linear operator

$$(P_k f)(\mathbf{x}) = \sum_{\mathbf{j} \in \mathbf{Z}^s} (\wedge_k F)(\mathbf{j}) \phi(\mathbf{x} - \mathbf{j})$$

on $C(\mathbf{R}^s)$. Here and throughout, for a sequence $A = \{a_{\mathbf{j}}\}$, we use the notation $A(\mathbf{j}) = a_{\mathbf{j}}$. Note that $\wedge_k F$ is a symbolic approximation of the "interpolation operator" C. In addition, for any $F = \{f(\mathbf{j})\}$ where $f \in C(\mathbf{R}^s)$ as above, we may define the linear functionals $\lambda_{k,\mathbf{j}}$ on $C(\mathbf{R}^s)$ by

$$\lambda_{k,\mathbf{j}}(f) = (\wedge_k F)(\mathbf{j}),$$

and this now yields the "quasi-interpolation" operators:

(8.35) $$(P_k^h f)(\mathbf{x}) = \sum_{\mathbf{j} \in \mathbf{Z}^s} \lambda_{k,\mathbf{j}}(\sigma_{h^{-1}} f) \phi(\frac{1}{h}\mathbf{x} - \mathbf{j}).$$

That these operators are indeed quasi-interpolation operators for all large values of $k \in \mathbf{Z}_+$ can be seen from the following result.

THEOREM 8.8. *Let the commutator degree of ϕ be $n \in \mathbf{Z}_+$. Then $P_k^h p = p$ for all $p \in \pi_n^s$ and all $k \geq n$. Furthermore, if, in addition, ϕ is symmetric in the sense that $\phi(\mathbf{j}) = \phi(-\mathbf{j})$ for all $\mathbf{j} \in \mathbf{Z}^s$, then $P_k^h p = p$ for all $p \in \pi_n^s$ and all $k \geq \frac{1}{2}(n-1)$.*

Since (8.15) is satisfied, we have

$$\sum_{\mathbf{j} \in \mathbf{Z}^s} M(\mathbf{j}) = 0$$

where the sequence M has compact support. In other words, M is a (finite) linear combination of certain first differences, and this implies that

$$M^\ell P_h = 0$$

for all $\ell \geq n+1$ and $P_h = \{p(h\mathbf{m})\}$ where $p \in \pi_n^s$. Hence, for $k \geq n$ and $p \in \pi_n^s$, we have

$$\begin{aligned}(P_k^h p)(h\mathbf{m}) &= \sum_{\mathbf{j} \in \mathbf{Z}^s} (\wedge_k P_h)(\mathbf{j}) \phi(\mathbf{m}-\mathbf{j}) \\ &= (\wedge_k P_h \Phi)(\mathbf{m}) = (\wedge_k \Phi P_h)(\mathbf{m}) \\ &= ((I + \cdots + M^k)(I - M) P_k)(\mathbf{m}) \\ &= (I - M^{k+1}) P_h(\mathbf{m}) = P_h(\mathbf{m}) = p(h\mathbf{m})\end{aligned}$$

for all $\mathbf{m} \in \mathbf{Z}^s$. On the other hand, since $p \in \pi_n^s$ and the commutator of ϕ has degree n, it follows from the definition of the commutator that $P_k^h p$ is also in π_n^s. This means that the two polynomials $P_k^h p$ and p agree at all $h\mathbf{m}$, where $\mathbf{m} \in \mathbf{Z}^s$, and hence, must be identical. If, in addition, ϕ is symmetric, then M is a (finite) linear combination of certain second central differences, and this implies that

$$M^{k+1} P_h = 0$$

for all $k \geq \frac{1}{2}(n-1)$ and $P_h = \{p(h\mathbf{m})\}$, where $p \in \pi_n^s$. By using the same argument as above, we have completed the proof of the theorem.

Remarks.
(1) The same argument also shows that if the commutator indicator set $\Gamma = \Gamma_\phi$ lies in the set $\{\alpha \in \mathbf{Z}_+^s : |\alpha| \leq N\}$, then $P_k^h p = p$ for all $p \in \pi_\Gamma^s$ and all $k \geq N$. Of course, this conclusion holds for all $k \geq \frac{1}{2}(N-1)$ if ϕ is

symmetric. Hence, for all appropriate values of k, depending on whether ϕ is symmetric or not, the quasi-interpolation operators P_k^h satisfy the same approximation properties (8.31) and (8.32) as Q_Γ^h in Theorem 8.7.

(2) Although the linear functionals $\lambda_{k\mathbf{l}}$ have fairly large supports, the computational procedure of the corresponding quasi-interpolation operators P_k^h is very efficient (cf. Appendix). In the special case where ϕ is a box spline, Dahmen and Micchelli [97] consider some of the questions on quasi-interpolants whose defining linear functionals have somewhat smaller supports.

(3) We will see in the next chapter that in some situations the sequence $\{P_k^h\}$ converges to an interpolation operator that also provides the optimal order of approximation from $S(\phi)$.

(4) If the compactly supported continuous function ϕ is a piecewise polynomial, then we believe that the results in this and the following chapters always hold without any assumption on the Fourier transform of ϕ. In any case, any condition that guarantees everywhere pointwise convergence of the Fourier series with Fourier coefficients $\hat{\phi}(2\pi\mathbf{j})$ is also sufficient. Of course, the results in both Chapters 8 and 9 can be easily generalized to the setting where ϕ is not assumed to have compact support but decays sufficiently fast to zero at infinity. Hence, the methods and results here should have useful applications to constructing quasi-interpolants and interpolants when a more general function ϕ, such as a certain linear combination of translates of some radial function to be discussed next, is used.

CHAPTER 9

Multivariate Interpolation

Although Hermite interpolation by multivariate splines in $S_d^r(\Delta)$ can be easily accomplished by using vertex splines if $d \geq 2^s r + 1$ or generalized vertex splines if d is smaller but Δ is subject to certain symmetric refinement (see Chapter 6), the subject of multivariate Lagrange interpolation from $S_d^r(\Delta)$ is very much underdeveloped. The most difficult problem is, perhaps, giving some characterization of the location of sample points so that the interpolation problem is *poised*; by this we mean that for any data set whose cardinality agrees with the dimension of some subspace S_L of $S_d^r(\Delta)$, there is one and only one function from S_L that interpolates the given data. Even the special case where $S_L = S_1^0(\Delta)$ and Δ is a (regular) triangulation of some region Ω in $\mathbf{R}^2$ does not have a satisfactory solution (see §9.2). Hence, most recent efforts on interpolation by multivariate (polynomial) splines only dealt with gridded data; that is, data taken at sample points which lie on the vertices of a parallelepiped grid partition. We will consider two points of view in this chapter: the problem of cardinal interpolation where the data are taken on all of $\mathbf{Z}^s$, and the problem of scaled cardinal interpolation where the order of approximation by the interpolants from $S(\phi)$, say, is required to be optimal.

We remark, however, that a "global approach" by means of some "radial function" or certain linear combinations of translates of a radial function has recently been considered for interpolating scattered data with some success. Interpolation by a radial function may be interpreted as follows: for a given set of data $(\mathbf{x}^i, f_i)$ where $\mathbf{x}^i \in \mathbf{R}^s, f_i \in \mathbf{R}, 1 \leq i \leq N$, the problem is to determine $\{a_i\}$ and $p \in \pi_{m-1}^s$, with m depending on a given function $g: \mathbf{R}_+ \to \mathbf{R}$, such that

$$G(\mathbf{x}^i) = f_i, \quad i = 1, \cdots, N$$

where

$$G(\mathbf{x}) = \sum_{i=1}^{N} \alpha_i g(|\mathbf{x} - \mathbf{x}^i|^2) + p(\mathbf{x})$$

and

$$\sum_{i=1}^{N} \alpha_i\, q(\mathbf{x}^i) = 0, \quad q \in \pi_{m-1}^s.$$

If the function g is chosen to be

$$g(t) = \begin{cases} t^{m-s/2} \log t & \text{for even } s \\ t^{m-s/2} & \text{for odd } s, \end{cases}$$

then G is called a thin plate spline (cf. Duchon [101] and Meinguet [155]). If we set

$$g(t) = (h^2 + t)^{-1/2}$$

with $s = 2$ and $m = 0$, then G is called the inverse (or reciprocal) Hardy multiquadric (cf. Hardy [127]). In general, g may be chosen to be a conditionally nonnegative definite C^∞ function on $(0, \infty)$ of order m; that is,

$$(-1)^j g^{(j)}(t) \geq 0\ ,\quad t > 0 \text{ and } j = m, m+1, \cdots$$

(cf. Micchelli [158]). Recent advances include the papers by Dyn, Goodman, and Micchelli [103], Jackson [136], Madych and Nelson [152], Micchelli [158], and Powell [163]. Since our monograph specializes in multivariate polynomial spline functions, we are not going into details here but refer the reader to the recent survey articles by Dyn [102] and Franke [110].

9.1. Interpolation by polynomials. In this section we will consider the problem of characterizing the location of sample points, which we will call *nodes* for short, that guarantees poisedness of Lagrange interpolation by multivariate polynomials. Since the dimension of the polynomial space π_n^s is

$$N_n^s = \binom{n+s}{s}$$

we will consider a set of distinct nodes of cardinality N_n^s. This set will be said to *admit unique Lagrange interpolation from* π_n^s if for any given data, there is one and only one polynomial from π_n^s that interpolates the data at these nodes. The problem of characterizing nodes that admit unique Lagrange interpolation is very old and many papers have been written on this topic during the last two decades. We mention specifically the paper by Chung and Yao [78] because it has inspired the research of several of the other papers. In a previous CBMS-NSF conference publication, Cheney (cf. [46]) also discussed multivariate Lagrange polynomial interpolation briefly and suggested a computational strategy. In order not to repeat the same approach and also to discuss the connection with Hermite and Birkhoff interpolations, we will follow the approach that Chui and Lai [68] take in obtaining the determinants of the coefficient matrices in product form so that their recurrence relationships allow us to consider coalescence of nodes along a line which yields Hermite interpolation, coalescence of lines on the same plane which gives rise to Birkhoff interpolation, and so on.

We will first consider the bivariate setting, and for simplicity, we use the notation
$$N_n = N_n^2.$$
Let $\widehat{X}_n = \{\mathbf{x}^1, \ldots, \mathbf{x}^{N_n}\}$ be a set of nodes. The following procedure for placing these nodes, which will be called a *node configuration*, is important.

Node Configuration L. *There exist $n+1$ distinct lines $\gamma_0, \cdots, \gamma_n$ such that $n+1$ of the distinct nodes $\mathbf{x}^1, \cdots, \mathbf{x}^{N_n}$ in $\widehat{X}_n$ lie on γ_n, n nodes lie on $\gamma_{n-1} \backslash \gamma_n, \cdots$, one node lies on $\gamma_0 \backslash (\gamma_1 \cup \cdots \cup \gamma_n)$. By relabeling the nodes if necessary, assume that $\mathbf{x}^{N_{j-1}+1}, \cdots, \mathbf{x}^{N_j} \in \gamma_j \backslash (\gamma_{j+1} \cup \cdots \cup \gamma_n)$ for $j = 0, \cdots, n-1$ where $N_{-1} = 0$, and $\mathbf{x}^{N_{n-1}+1}, \cdots, \mathbf{x}^{N_n} \in \gamma_n$.*

It is well known that if $\widehat{X}_n$ satisfies Node Configuration L, then it admits unique Lagrange interpolation (from π_n^2). In the following, to prove this fact, we will even give the determinant of the coefficient matrix

(9.1) $$[\phi_1 \cdots \phi_{N_n}]$$

with $\phi_i = [(\mathbf{x}^i)^{\mathbf{j}}]^T$, $|\mathbf{j}| \leq n$, as its ith column.

Let $\mathbf{x} = (x, y)$ be the variables in $\mathbf{R}^2$ and ψ_m denote the monomials
$$\psi_m(\mathbf{x}) = x^{k-m+N_{k-1}+1} y^{m-N_{k-1}-1}$$
for $N_{k-1} < m \leq N_k$ and $k = 0, 1, \cdots, n$. The determinant of the matrix in (9.1) is a bivariate Vandermonde determinant and we will denote it by

$$VD_n \begin{pmatrix} \psi_1, \cdots, \psi_{N_n} \\ \mathbf{x}^1, \cdots, \mathbf{x}^{N_n} \end{pmatrix} = \det[\phi_1 \cdots \phi_{N_n}]$$

where $\phi_i = [\psi_1(\mathbf{x}^i) \cdots \psi_{N_n}(\mathbf{x}^i)]^T$.

THEOREM 9.1. *Let $\widehat{X}_n \subset \mathbf{R}^2$ satisfy Node Configuration L. Then*

(9.2)
$$VD_n \begin{pmatrix} \psi_1, \cdots, \psi_{N_n} \\ \mathbf{x}^1, \cdots, \mathbf{x}^{N_n} \end{pmatrix}$$
$$= c \prod_{k=1}^{n} \prod_{N_{k-1} < j < i \leq N_k} |\mathbf{x}^i - \mathbf{x}^j| \prod_{p=1}^{n} \prod_{q=1}^{N_{p-1}} d(\mathbf{x}^q, \gamma_p)$$

where $c = 1$ or -1 and $d(\mathbf{x}^q, \gamma_p)$ denotes the distance from $\mathbf{x}^q$ to γ_p. In particular, $\widehat{X}_n$ admits unique Lagrange interpolation from π_n^2.

In Chui and Lai [68], (9.2) is established by mathematical induction, and for this purpose, the following recurrence relationship is obtained.

LEMMA 9.1. *Let*
$$\mathbf{x}^{N_{n-1}+j} = (x_1^{N_{n-1}+j}, a)$$

$j = 1, \cdots, n+1$, *be distinct points that lie on some horizontal line $y = a$. Then*

(9.3)
$$VD_n \begin{pmatrix} \psi_1 \cdots \psi_{N_n} \\ \mathbf{x}^1 \cdots \mathbf{x}^{N_n} \end{pmatrix}$$
$$= c \prod_{k=1}^{N_{n-1}} (x_2^k - a) \prod_{1 \leq i < j \leq n+1} (x_1^{N_{n-1}+j} - x_1^{N_{n-1}+i}) VD_{n-1} \begin{pmatrix} \psi_1 \cdots \psi_{N_{n-1}} \\ \mathbf{x}^1 \cdots \mathbf{x}^{N_{n-1}} \end{pmatrix}$$

where $c = 1$ or -1.

In the above formula, note that $N_{n-1} + n + 1 = N_n$. In addition, we have used the line $y = a$ for γ_n to simplify notation. It should be noted that a rotation only involves multiplying the coefficient matrix by a matrix with determinant equal to 1 or -1.

The importance of formula (9.3) is that the first components of all the nodes on γ_n are singled out as a Vandermonde product. This allows us to divide a factor of this product to both sides of (9.3) and take limits, yielding directional derivatives along γ_n in the coefficient matrix. By applying formula (9.3) consecutively for $n-1, n-2, \cdots, 2$ and using the same procedure in each step (with an appropriate rotation), we can show that coalescence of nodes along each γ_j only changes Lagrange interpolation to Hermite interpolation but does not destroy the poisedness of the interpolation problem. The node configuration can be described as follows.

Node Configuration H. *There exist $n+1$ distinct lines $\gamma_0, \cdots, \gamma_n$ such that, by relabeling the nodes $\mathbf{x}^1, \cdots, \mathbf{x}^{N_n}$ in $\widehat{X}_n$, if necessary,*

$$\mathbf{x}^{N_{j-1}+1}, \cdots, \mathbf{x}^{N_j} \in \gamma_j \backslash (\gamma_{j+1} \cup \cdots \cup \gamma_n)$$

for $j = 0, \cdots, n-1$,

$$\mathbf{x}^{N_{n-1}+1}, \cdots, \mathbf{x}^{N_n} \in \gamma_n,$$

and

$$\mathbf{x}^{N_{j-1}+1}, \cdots, \mathbf{x}^{N_j} = \underbrace{\mathbf{y}^{j1}, \cdots, \mathbf{y}^{j1}}_{\ell_{j1}}, \cdots, \underbrace{\mathbf{y}^{jk_j}, \cdots, \mathbf{y}^{jk_j}}_{\ell_{jk_j}}$$

with $\ell_{j1} + \cdots + \ell_{jk_j} = N_j - N_{j-1} = j + 1$, $j = 0, \cdots, n$.

By using the procedure outlined above, we can even obtain the determinant of the coefficient matrix for Hermite interpolation. In the following, we will use the notation HD_n to indicate that directional derivatives must be taken in the coefficient matrix when coalescent nodes occur.

THEOREM 9.2. *Let $\widehat{X}_n$ satisfy Node Configuration H. Then the determinant of the coefficient matrix of the corresponding Hermite interpolation problem is given by*

$$(9.4) \qquad HD_n \begin{pmatrix} \psi_1 \cdots \psi_{N_n} \\ \mathbf{x}^1 \cdots \mathbf{x}^{N_n} \end{pmatrix}$$

$$= c \prod_{j=1}^{n} \prod_{1 \le s < t \le k_j} |\mathbf{y}^{js} - \mathbf{y}^{jt}|^{\ell_{js}\ell_{jt}} \prod_{u=1}^{k_j} \prod_{\rho=1}^{\ell_{ju}-1} \rho! \prod_{v=1}^{n} \prod_{r=0}^{v-1} \prod_{w=1}^{k_r} (d(\mathbf{y}^{rw}, \gamma_v))^{\ell_{rw}}$$

where $c = 1$ or -1. In particular, $\widehat{X}_n$ admits unique Hermite interpolation from π_n^2.

We next consider coalescence of the lines γ_j. To introduce this idea, let us study a simple example.

Example 9.1. Let $\widehat{X}_1 = \{\mathbf{x}^1, \mathbf{x}^2, \mathbf{x}^3\}$ with $\mathbf{x}^i = (d_i, 0), d_1 < d_2 < d_3$. Obviously, $\widehat{X}_1$ does not admit unique Lagrange interpolation from π_1^2. Indeed, any linear polynomial

$$p(x, y) = a + bx + cy$$

satisfies $p(\mathbf{x}^i) = a + bd_i, i = 1, 2, 3$, so that the coefficient c is not effected. One method to determine c is to take the derivative with respect to y, which is the *directional derivative* in the direction normal to the line $\gamma_1 \colon y = 0$ (that contains $\mathbf{x}^1, \mathbf{x}^2, \mathbf{x}^3$). That is, we consider the Birkhoff interpolation problem

$$\begin{cases} (f-p)(\mathbf{x}^i) = 0, & i = 1, 2 \\ \dfrac{\partial}{\partial y}(f-p)(\mathbf{x}^3) = 0 \end{cases}$$

which is now poised. Taking the derivative with respect to y corresponds to interpolating at $\mathbf{x}^1, \mathbf{x}^2, \mathbf{x}^3 + (0, \varepsilon)$ and (after subtracting by a zero term and dividing by ε) taking the limit as $\varepsilon \to 0$; this is equivalent to bringing the line $y = \varepsilon$ toward the x-axis.

The above example suggests the following node placement.

Node Configuration B. There exist lines $\gamma_0, \cdots, \gamma_n$ with

$$\gamma_0, \cdots, \gamma_n = \underbrace{\beta_1, \cdots, \beta_1}_{m_1}, \cdots, \underbrace{\beta_d, \cdots, \beta_d}_{m_d},$$

$m_1 + \cdots + m_d = n+1$ and $\beta_1, \cdots, \beta_d$ distinct, such that $\mathbf{x}^{N_{j-1}+1}, \cdots, \mathbf{x}^{N_j}$ lie on γ_j but not on those of $\gamma_{j+1}, \cdots, \gamma_n$ that are distinct from γ_j, $j = 0, \ldots, n$.

For the nodes $\mathbf{x}^{N_{n-1}+1}, \cdots, \mathbf{x}^{N_n}$ on $\beta_d = \gamma_n$ (which is the last β_d), we assign Lagrange interpolation:

$$(f-p)(\mathbf{x}^i) = 0, \ i = N_{n-1}+1, \cdots, N_n,$$

for the nodes on $\beta_d = \gamma_{n-1}$ (the next to the last β_d), we assign interpolation of the first directional derivative along the line orthogonal to β_d, and so on; and finally, to the last set of nodes on the first $\beta_d = \gamma_{n-m_d+1}$, we assign interpolation of the $(m_d - 1)$st-directional derivatives along the same normal. The same Birkhoff interpolation scheme is performed at the nodes on the other lines $\beta_{d-1}, \cdots, \beta_1$. We denote the determinant of the coefficient matrix of this interpolation problem by

$$BD_n \begin{pmatrix} \psi_1, \cdots, \psi_{N_n} \\ \mathbf{x}^1, \cdots, \mathbf{x}^{N_n} \end{pmatrix}$$

which is unique up to a multiplicative constant $c = 1$ or -1. Using the formula (9.2) in Theorem 9.1 and the procedure suggested by the above example, we have the following theorem.

THEOREM 9.3. *Let $\widehat{X}_n$ satisfy Node Configuration B. Then the determinant of the coefficient matrix of the Birkhoff interpolation problem described above is given by*

$$(9.5) \quad BD_n \begin{pmatrix} \psi_1, \cdots \psi_{N_n} \\ \mathbf{x}^1, \cdots, \mathbf{x}^{N_n} \end{pmatrix}$$

$$= c \prod_{k=1}^{n} \prod_{N_{k-1} < j < i \leq N_k} |\mathbf{x}^i - \mathbf{x}^j| \prod_{p=2}^{d} \prod_{\mathbf{x}^q \in B_p} d(\mathbf{x}^q, \beta_p)$$

$$\times \prod_{u=1}^{d} \prod_{v=1}^{m_u - 1} (v!)^{m_1 + \cdots + m_u - v}$$

where $B_p = \beta_1 \cup \cdots \cup \beta_{p-1}$ and $c = 1$ or -1. Consequently, $\widehat{X}_n$ admits unique Birkhoff interpolation from π_n^2 as described above.

All the above results can be easily extended to $\mathbf{R}^s$ for $s > 2$ by an induction procedure. Since the notation would have to be quite complicated, we only consider Lagrange interpolation but will formulate the determinant of the coefficient matrix so that a procedure analogous to the one in $\mathbf{R}^2$ can be applied.

Node Configuration L in $\mathbf{R}^s$. Let $\widehat{X} = \{\mathbf{x}^i : i = 1, \cdots, N_n^s\}$ be a set of distinct points in $\mathbf{R}^s$ such that, after relabeling, if necessary, there exist $n + 1$ hyperplanes $K_i^s, i = 0, \cdots n$, with

$$\mathbf{x}^{N_{n-1}^s + 1}, \cdots, \mathbf{x}^{N_n^s} \in K_n^s$$

and

$$\mathbf{x}^{N_{j-1}^s + 1}, \cdots, \mathbf{x}^{N_j^s} \in K_j^s \backslash (K_{j+1}^s \cup \cdots \cup K_n^s)$$

for $j = 0, \cdots, n - 1$, and that each set of points

$$\{\mathbf{x}^{N_{j-1}^s + 1}, \cdots, \mathbf{x}^{N_j^s}\},$$

where $0 \leq j \leq n$, considered as points in $\mathbf{R}^{s-1}$, satisfies Node Configuration L in $\mathbf{R}^{s-1}$.

THEOREM 9.4. *Let $\widehat{X} = \{\mathbf{x}^i : i = 1, \cdots, N_n^s\}$ satisfy Node Configuration L in $\mathbf{R}^s$. Then*

$$(9.6) \quad VD_n^s \begin{pmatrix} \psi_1^s, \cdots, \psi_{N_n^s}^s \\ \mathbf{x}^1, \cdots, \mathbf{x}^{N_n^s} \end{pmatrix}$$

$$= c \prod_{\ell=1}^{n} \prod_{i=1}^{N_{\ell-1}^s} d(\mathbf{x}^i, K_\ell^s) \prod_{p=1}^{n} VD_p^{s-1} \left(\begin{matrix} \psi_1^{s-1}, \cdots, \psi_{N_p^s-1}^{s-1} \\ \mathbf{x}^{N_{p-1}^s+1}, \cdots, \mathbf{x}^{N_p^s} \end{matrix} \right) \bigg|_{K_p^s}$$

where $c = 1$ or -1 and $VD_q^2 = VD_q$. Consequently, $\widehat{X}$ admits unique Lagrange interpolation from π_n^s.

9.2. Lagrange interpolation by multivariate splines. The title of this section is, perhaps, somewhat misleading since even the simplest situation of interpolation by piecewise linear polynomials in two variables still requires further investigation. The objective of this section is to introduce to the reader a very important research area that requires new ideas and techniques for any direction of development. The problem is *characterization of the sample points* (or *nodes*) *that admit unique Lagrange interpolation from a subspace of a given multivariate spline space* $S_d^r(\Delta, \Omega), \Omega \subseteq \mathbf{R}^s$, *where the number of nodes is given by the dimension of the subspace.*

We are interested in any subspace S_L with dimension n and having a basis which consists of locally supported functions $B_1, \cdots, B_n$, say. Let $\widehat{X} = \{\mathbf{x}^1, \cdots, \mathbf{x}^n\}$ be a set of distinct nodes. The problem is to study node configurations for $\widehat{X}$ in Ω such that for any given data $\{f_1, \cdots, f_n\}$ there exists one and only one function $s \in S_L$ such that

$$s(\mathbf{x}^i) = f_i \quad , \quad i = 1, \cdots, n.$$

Recall that for univariate splines, this problem is completely solved in Schoenberg and Whitney [182] (see Theorem 1.6 in §1.4). In the following, we will demonstrate with two examples that there does not seem to be a simple extension of the Schoenberg–Whitney theorem to the multivariate setting.

Let Δ be any regular but otherwise arbitrary triangulation of some (polygonal) region Ω in $\mathbf{R}^2$; that is, the conditions (i), (ii), (iii) in §4.3 are satisfied. Then if $\{A_1, \cdots, A_V\}$ denotes the set of all vertices including both the interior and boundary ones of Δ, we know that the dimension of the space $S_1^0(\Delta)$ of piecewise linear polynomials in $C(\Omega)$ is the number V of vertices in Δ, and a basis is given by

$$\{s_1, \cdots, s_V\}$$

defined uniquely by $s_i(A_j) = \delta_{ij}, 1 \leq i, j \leq V$. Of course each s_i is a minimally supported function in $S_1^0(\Delta)$. It is usually called a Courant (hat) function.

Example 9.2. Let $\Omega = [-1,1]^2$ and Δ be defined by drawing in the two diagonals. Denote the five vertices by $A_1, \cdots, A_5$ as shown in Figure 9.1. Then any set $\widehat{X} = \{\mathbf{x}^1, \cdots, \mathbf{x}^5\}$ of five points that lie on a level curve of $z = s_5(\mathbf{x})$, say the boundary of $[-\frac{1}{2}, \frac{1}{2}]^2$, does not admit unique Lagrange interpolation from $S_1^0(\Delta)$.

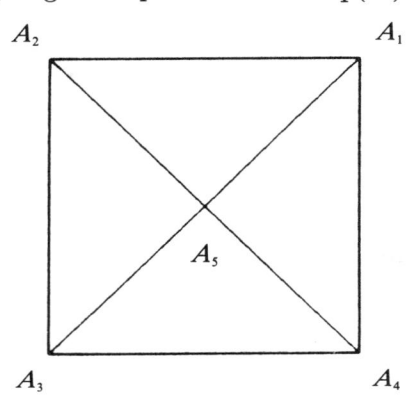

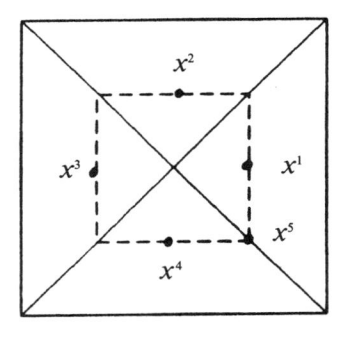

FIG. 9.1

Indeed, we have:

$$s_5(\mathbf{x}^i) = \frac{1}{2}(s_1 + \cdots + s_5)(\mathbf{x}^i) = \frac{1}{2}$$

for $i = 1, \cdots, 5$. In Figure 9.1 we have assigned

$$\mathbf{x}^1 = (\frac{1}{2}, 0), \quad \mathbf{x}^2 = (0, \frac{1}{2}), \quad \mathbf{x}^3 = (-\frac{1}{2}, 0),$$
$$\mathbf{x}^4 = (0, -\frac{1}{2}), \quad \mathbf{x}^5 = (\frac{1}{2}, -\frac{1}{2})$$

so that $\mathbf{x}^i$ lies in the interior of supp s_i for each i. Note that not only is the Schoenberg–Whitney condition satisfied, but also the intersection of any two supports contains an $\mathbf{x}^i$ in its interior.

Example 9.3. Consider the same Ω and partition Δ as in the above example. Place any three noncolinear points $\mathbf{x}^1, \mathbf{x}^2, \mathbf{x}^3$ in the triangle $A_3 A_4 A_5$. Also, place $\mathbf{x}^4$ on the positive y-axis with both $\mathbf{x}^4$ and $\mathbf{x}^5$ in the triangle $A_1 A_2 A_5$. Then the set $\widehat{X} = \{\mathbf{x}^1, \cdots, \mathbf{x}^5\}$ admits unique Lagrange interpolation from $S_1^0(\Delta)$ if and only if $\mathbf{x}^5$ does not lie on the y-axis. See Figure 9.2 below.

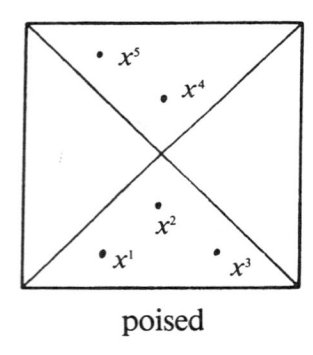

138   CHAPTER 9

                    poised 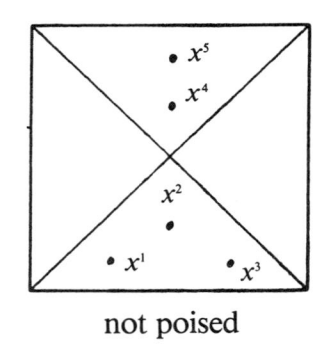   not poised
                                    FIG. 9.2

In view of the above examples, in developing any node configuration algorithm, it is convenient to tag on the following conditions.

(C1) At least one triangular subregion (which will be called a cell) contains three nodes in its closure. If the closure of a cell contains three nodes, they are not colinear. No cell contains more than three nodes in its closure.

(C2) If exactly two nodes are placed in the closure of a cell, they are not colinear with any vertex of this cell.

We remark, however, that these conditions are not necessary. In the following, we will give an algorithm for constructing a fairly general node configuration that guarantees unique Lagrange interpolation from $S_1^0(\Delta)$. We need the following terminology to facilitate our presentation. Throughout, $\widehat{X} = \{\mathbf{x}^1, \cdots, \mathbf{x}^V\}$ will denote a set of nodes, and Δ a regular but arbitrary triangulation of $\Omega \subset \mathbf{R}^2$ with V vertices, including both the interior and boundary ones.

(1) The closure of the support of a Courant function s_P with $s_P(P) = 1$ is called a *unit* relative to P and is denoted by U_P.

(2) If the total number of nodes in $\widehat{X}$ that have been placed on T, where T may be the closure of a cell or the union of a finite number of units, is the same as the number of vertices of Δ that lie in T, we say that the vertices of T and T itself have been labeled.

(3) Let U_P be a unit relative to a vertex P, and T be either the closure of a cell or the union of a finite number of units such that P is also a (boundary or interior) vertex of T. Suppose that all vertices of T have been labeled but U_P has not been completely labeled (and this means that not all vertices of U_P have been labeled). Let $L_1, \cdots, L_n$ be the edges with P as a common end-point such that their other end-points are also vertices of Δ that lie in T. Similarly, let $\Omega_1, \cdots, \Omega_m$ $(m < n)$

be closures of the triangular cells sharing P as a common vertex such that the other two vertices of each of these triangles are also vertices of $\triangle$ that lie in T. Then each of the components of

$$U_P \backslash (T \cup L_1 \cup \cdots \cup L_n \cup \Omega_1 \cup \cdots \cup \Omega_m)$$

is called an *unlabeled component* of $U_P \backslash T$.

(4) Let C_P be an *unlabeled component* of $U_P \backslash T$. If C_P has at least one boundary edge with one end-point at P and the other end-point (vertex) unlabeled, C_P is called an *unlabeled boundary component* of $U_P \backslash T$. Otherwise it is called an *unlabeled interior component* of $U_P \backslash T$.

The following algorithm is in Chui, He, and Wang [61].

Select any cell T^* with a vertex P_1 which may be either a boundary or interior vertex of $\triangle$. Place any three noncolinear nodes on the closure of this cell, and consider the unlabeled component(s) of $U_{P_1} \backslash T^*$. (Note that if P_1 is a boundary vertex of $\triangle$, $U_{P_1} \backslash T^*$ may have two unlabeled boundary components, and if P_1 is an interior vertex of $\triangle$, it has only one unlabeled interior component.) *Let C_{P_1} be an unlabeled component with n_1 unlabeled vertices, say. Denote the closure of the cells of C_{P_1} by $T_1, \cdots, T_{m_1}$ in the consecutive order starting from T_1 which is adjacent to T^* (in either a clockwise or counterclockwise direction if P_1 is an interior vertex), where $m_1 = n_1$ if C_{P_1} is a boundary component and $m_1 = n_1 + 1$ if it is an interior component. Place nodes on this component according to one of the following rules:*

(i) *Number of nodes on $T_1 \cup \cdots \cup T_j = j$ for $= 1, 3, \cdots, 2[n_1/2] - 1$, and total number of nodes on C_{P_1} is n_1, such that conditions* (C1) *and* (C2) *are satisfied, or*

(ii) *Number of nodes on $T_1 \cup \cdots \cup T_j = j$ for $j = 2, 4, \cdots, 2[n_1/2]$, and total number of nodes on C_{P_1} is n_1, such that conditions* (C1) *and* (C2) *are satisfied.*

If $U_{P_1} \backslash T^*$ has two components, apply the same procedure described above to each component. (It is now clear that the unit U_{P_1} relative to the vertex P_1 is labeled.)

Let $P_1, \cdots, P_r$ be distinct vertices, arbitrarily chosen, such that the corresponding units $U_i = U_{P_i}$ satisfy:

(a) *Number of vertices of $\triangle$ in $U_1 \cup \cdots \cup U_{r-1} < V$,*

(b) *Number of vertices of $\triangle$ in $U_1 \cup \cdots \cup U_r = V$, and*

(c) *$(U_1 \cup \cdots \cup U_i) \cap U_{i+1}$ contains at least one triangular cell, $i = 1, 2, \cdots, r - 1$.*

Suppose that the union

$$T_i = U_1 \cup \cdots \cup U_i$$

has been labeled. Then we place nodes on each unlabeled component C_{i+1} of $U_{i+1} \setminus T_i$ using either rule (i) *or rule* (ii) *(where n_1 is replaced by n_i, the number of unlabeled vertices of C_{i+1}, and C_{P_1} is replaced by C_{i+1}).*

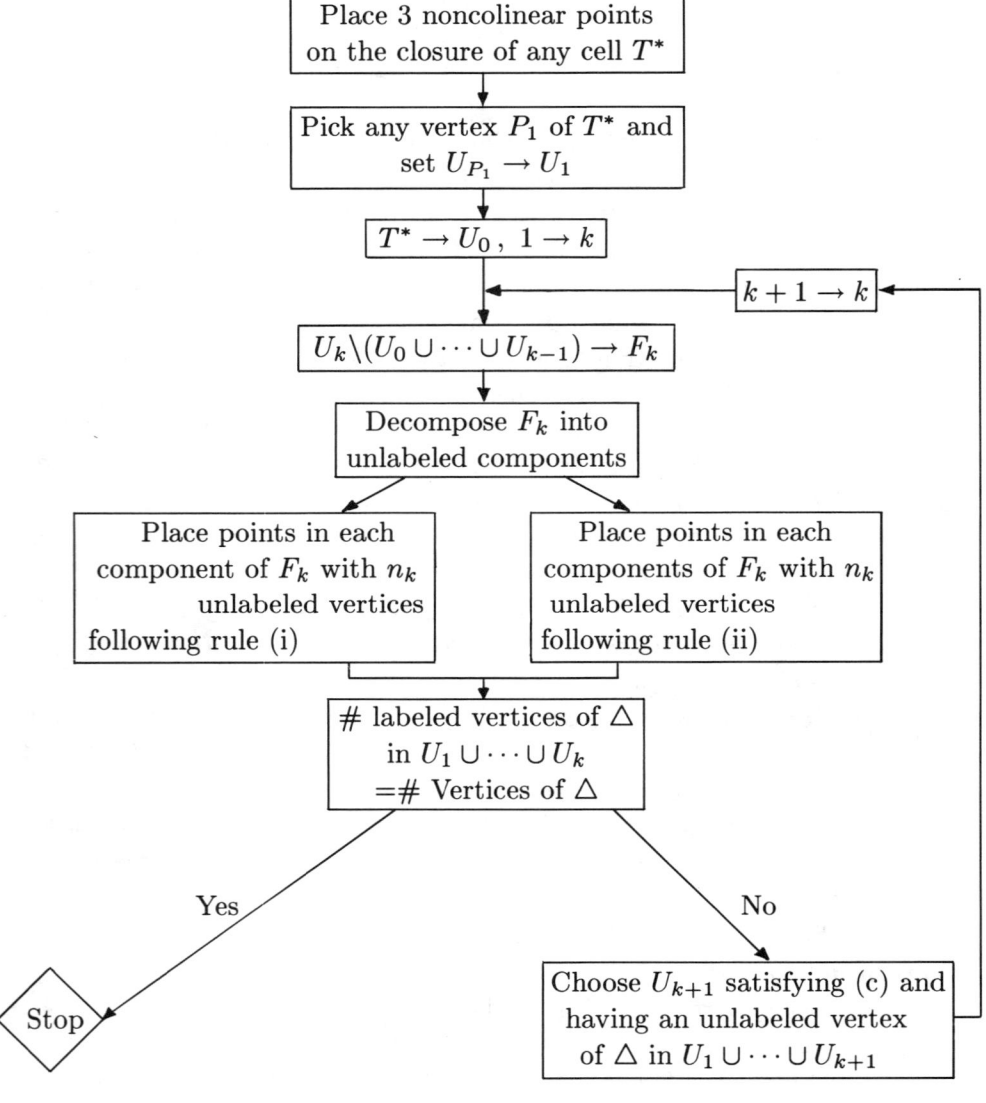

```
┌─────────────────────────────────┐
│ Place 3 noncolinear points      │
│ on the closure of any cell T*   │
└─────────────────────────────────┘
                │
                ▼
┌─────────────────────────────────┐
│ Pick any vertex P₁ of T* and    │
│ set U_{P_1} → U_1               │
└─────────────────────────────────┘
                │
                ▼
    ┌──────────────────────┐
    │ T* → U_0, 1 → k      │◄──────── k+1 → k ◄──┐
    └──────────────────────┘                     │
                │                                │
                ▼                                │
    ┌──────────────────────────────┐             │
    │ U_k\(U_0 ∪ ··· ∪ U_{k-1}) → F_k │          │
    └──────────────────────────────┘             │
                │                                │
                ▼                                │
    ┌──────────────────────┐                     │
    │ Decompose F_k into   │                     │
    │ unlabeled components │                     │
    └──────────────────────┘                     │
          │         │                            │
          ▼         ▼                            │
 ┌─────────────┐ ┌─────────────┐                 │
 │Place points │ │Place points │                 │
 │in each comp.│ │in each comps│                 │
 │of F_k with  │ │of F_k with  │                 │
 │n_k unlabeled│ │n_k unlabeled│                 │
 │vertices     │ │vertices     │                 │
 │following(i) │ │following(ii)│                 │
 └─────────────┘ └─────────────┘                 │
          │         │                            │
          └────┬────┘                            │
               ▼                                 │
    ┌──────────────────────────────┐             │
    │ # labeled vertices of △      │             │
    │ in U_1 ∪ ··· ∪ U_k           │             │
    │ = # Vertices of △            │             │
    └──────────────────────────────┘             │
        Yes  /        \  No                      │
            /          \                         │
           ▼            ▼                        │
        ◇Stop◇   ┌──────────────────────────┐    │
                 │ Choose U_{k+1} satisfying│    │
                 │ (c) and having an        │────┘
                 │ unlabeled vertex of △    │
                 │ in U_1 ∪ ··· ∪ U_{k+1}   │
                 └──────────────────────────┘
```

FIG. 9.3

Apply this procedure for $i = 1, 2, \cdots, r-1$. (Note that all units have now been labeled.) Let $\widehat{X}$ be the set of all such nodes. (See the flow chart in Figure 9.3.)

Then we have the following result.

THEOREM 9.5. *The set $\widehat{X}$ of nodes selected according to the above algorithm admits unique Lagrange interpolation from $S_1^0(\Delta)$.*

To study Lagrange interpolation node configurations for bivariate C^1 quadratic splines, we must restrict our attention to a triangulation that allows locally supported functions. In Chui and He [58], several node configurations that admit unique Lagrange interpolation from $S_2^1(\Delta_{MN}^{(2)})$ are given, where $\Delta_{MN}^{(2)}$ is a nonuniform (i.e., not necessarily uniform) type-2 triangulation of the rectangular region Ω_{MN}.

9.3. Cardinal interpolation with nonsingular ϕ.

The remaining material in this chapter may be viewed as a continuation of Chapter 8, where the vector space $S(\phi)$ of ϕ-series

$$\sum_{\mathbf{j} \in \mathbf{Z}^s} c_{\mathbf{j}} \phi(\cdot - \mathbf{j})$$

or more precisely, the corresponding space of scaled translates,

$$S_h(\phi) = \{\sigma_h f : f \in S(\phi)\}, \quad h > 0,$$

was considered as an approximating space. We will mainly study interpolation from $S(\phi)$ with $\mathbf{Z}^s$ as the set of nodes. Interpolation from $S_h(\phi)$ at $h\mathbf{Z}^s$, $h > 0$, will be discussed in §9.5. Here and throughout, ϕ will always denote a compactly supported continuous function in $\mathbf{R}^s$ with $\{\hat{\phi}(2\pi \mathbf{j})\} \in \ell^1(\mathbf{Z}^s)$. It may be a box spline, a linear combination of several box splines, or minimally and quasi-minimally supported splines.

The problem of cardinal interpolation from $S(\phi)$ can be stated as follows: *For a given data sequence $F = \{f(\mathbf{j})\}$, $\mathbf{j} \in \mathbf{Z}^s$, determine a coefficient sequence $C = \{c_{\mathbf{j}}\}$, $\mathbf{j} \in \mathbf{Z}^s$, such that*

$$s_f(\mathbf{x}) = \sum_{\mathbf{j} \in \mathbf{Z}^s} c_{\mathbf{j}} \phi(\mathbf{x} - \mathbf{j})$$

from $S(\phi)$ satisfies

$$s_f(\mathbf{k}) = f(\mathbf{k}), \quad \mathbf{k} \in \mathbf{Z}^s,$$

or equivalently,

(9.7) $$C\Phi = F$$

where $\Phi = \{\phi(\mathbf{j})\}$ and the sequence product notation in (9.7) represents convolution on $\mathbf{Z}^s$ as already defined in (8.33).

Hence, the problem is to determine C that satisfies (9.7). Following the presentation in Chui, Jetter, and Ward [63], let us consider the operator

(9.8) $$L_\phi: \ell^p \to \ell^p, \quad 1 \le p \le \infty,$$

defined by
$$L_\phi C = C\Phi,$$

and we are interested in the problem of characterizing the compactly supported function ϕ so that L_ϕ has a (continuous) inverse. For this purpose, we introduce the trigonometric polynomial

$$\tilde{\phi}(\omega) = \sum_{\mathbf{j} \in \mathbf{Z}^s} \phi(\mathbf{j}) e^{-i\mathbf{j}\cdot\omega}$$

induced by ϕ. Note that $\tilde{\phi}$ is the Fourier transform of ϕ over $\mathbf{Z}^s$ and is also called the *discrete Fourier transform* of ϕ. It is the restriction of the *symbol*

$$\sigma_\phi(\mathbf{z}) = \sum_{\mathbf{j} \in \mathbf{Z}^s} \phi(j) \mathbf{z}^{-\mathbf{j}}$$

of ϕ on the unit polycircle, T^s: $\mathbf{z} = e^{i\omega}, \omega \in \mathbf{R}^s$. The following result can be proved by using standard techniques from Fourier transforms.

LEMMA 9.2. *The following statements are equivalent:*
(i) $L_\phi: \ell^2 \to \ell^2$ *has a continuous inverse.*
(ii) $L_\phi: \ell^p \to \ell^p$ *has a continuous inverse for all $p, 1 \le p \le \infty$.*
(iii) *The symbol σ_ϕ does not vanish anywhere on T^s.*

Remarks.

(1) If ϕ is symmetric on $\mathbf{Z}^s$; that is, $\phi(\mathbf{j}) = \phi(-\mathbf{j})$ for all $\mathbf{j} \in \mathbf{Z}^s$, then L_ϕ is a symmetric operator on ℓ^2, and condition (iii) above is equivalent to saying that the discrete Fourier transform $\tilde{\phi}$ is either strictly positive or strictly negative on $[-\pi, \pi]^s$.

(2) In de Boor, Höllig, and Riemenschneider [36], it is shown that condition (iii) is equivalent to the condition that L_ϕ is a one-one mapping

from ℓ^∞ onto itself. Hence, by an application of the Banach inverse theorem, it is clear that L_ϕ has a continuous inverse.

Returning to the cardinal interpolation problem, we say that the problem is ℓ^p-*solvable* from $S(\phi)$ if for any data sequence $F \in \ell^p$, there exists a unique $C \in \ell^p$ such that $C\Phi = F$. Hence, in view of Remark (2), Lemma 9.2 can be restated as follows:

THEOREM 9.6. *The following statements are equivalent:*
(i) *The cardinal interpolation problem from $S(\phi)$ is ℓ^2-solvable.*
(ii) *The cardinal interpolation problem from $S(\phi)$ is ℓ^p-solvable for all p, $1 \leq p \leq \infty$.*
(iii) *The symbol σ_ϕ does not vanish anywhere on T^s.*

If condition (iii) in the above theorem is satisfied, ϕ is said to be *nonsingular*. Hence, by Theorem 9.6, it is quite convenient to solve the cardinal interpolation problem by using a nonsingular ϕ. The following result is proved in de Boor, Höllig, and Riemenschneider [85].

THEOREM 9.7. *Let $M(\cdot|X_n)$ be a bivariate box spline with direction set $X_n \subset \mathbf{Z}^2 \setminus \{\mathbf{0}\}$. Then the symbol of $M(\cdot|X_n)$ is strictly positive on T^s if and only if the translates $M(\cdot - \mathbf{j}|X_n), \mathbf{j} \in \mathbf{Z}^2$, are linearly independent.*

Hence, any box spline M_{tuv} on the three-directional mesh guarantees that the corresponding cardinal interpolation is ℓ^p-solvable, but if a box spline M_{tuvw} on the four-directional mesh is used for ϕ, the problem is no longer ℓ^p-solvable (cf. Theorem 2.11). For box splines in $\mathbf{R}^s$ with $s > 2$, however, de Boor, Höllig, and Riemenschneider [37] have demonstrated with an example that, at least for large values of s, linear independence of $M(\cdot - \mathbf{j}|X_n)$ does not necessarily imply that the symbol of $M(\cdot|X_n)$ is nonvanishing on T^s. For singular ϕ, Chui, Diamond, and Raphael [54] introduce a method to construct cardinal interpolants from $S(\phi)$ where σ_ϕ may have isolated zeros or zero manifolds of codimension one on T^s. These results will be discussed in the next section.

In the rest of this section, we will restrict our attention to compactly supported continuous functions ϕ in $\mathbf{R}^s$ with $\{\hat{\phi}(2\pi \mathbf{j})\} \in \ell^1(\mathbf{Z}^s)$ that satisfy the following conditions:
(i) ϕ is normalized in the sense that

$$\sum_{\mathbf{j} \in \mathbf{Z}^s} \phi(\mathbf{j}) = 1,$$

(ii) $\phi(\mathbf{j}) \geq 0$ and $\phi(-\mathbf{j}) = \phi(\mathbf{j})$ for all $\mathbf{j} \in \mathbf{Z}^s$,
and

(iii) the symbol σ_ϕ of ϕ does not vanish on T^s; that is,

$$\tilde{\phi}(\omega) = \sum_{\mathbf{j} \in \mathbf{Z}^s} \phi(\mathbf{j}) e^{-i \mathbf{j} \cdot \omega} > 0$$

for all $\omega \in \mathbf{R}^s$.

As in §8.4, let $I = \{\delta_{\mathbf{j0}}\}$ denote the convolution identity. Then by Theorem 9.6, the cardinal interpolation problem from $S(\phi)$, with data sequence I, has a unique solution $C^* = \{c_\mathbf{j}^*\} \in \ell^p$, $1 \leq p \leq \infty$. Taking the discrete Fourier transform on both sides of

$$C^* \Phi = I$$

and using the assumption (iii) above, we have:

$$c_\mathbf{j}^* = \frac{1}{(2\pi)^s} \int_{(-\pi,\pi)^s} \tilde{C}^*(w) e^{i \mathbf{j} \cdot \omega} dw$$
$$= \frac{1}{(2\pi)^s} \int_{(-\pi,\pi)^s} \frac{e^{i \mathbf{j} \cdot \omega}}{\tilde{\phi}(\omega)} dw.$$

Hence, it follows that $\{c_\mathbf{j}^*\}$ decays exponentially as $|\mathbf{j}| \to \infty$, and that the corresponding solution

(9.9) $$L(\mathbf{x}) = \sum_{\mathbf{j} \in \mathbf{Z}^s} c_\mathbf{j}^* \phi(\mathbf{x} - \mathbf{j})$$

to the cardinal interpolation problem with data I also decays exponentially as $|\mathbf{x}| \to \infty$. The above discussion is a multivariate extension of a univariate cardinal interpolation result presented in a previous CBMS-NSF conference publication by Schoenberg (cf. [180]). For more information, the reader is referred to de Boor, Höllig, and Riemenschneider [36]. We will call $L(\mathbf{x})$ the *fundamental function* associated with ϕ. The following result can now be easily verified (cf. Chui, Jetter, and Ward [63]).

THEOREM 9.8. *Let ϕ satisfy the above conditions* (i) - (iii). *Then for any given data sequence* $F = \{f(\mathbf{j})\}$ *in* ℓ^p, $1 \leq p \leq \infty$, *the unique solution*

(9.10) $$s(\mathbf{x}) = \sum_{\mathbf{j} \in \mathbf{Z}^s} c_\mathbf{j} \phi(\mathbf{x} - \mathbf{j}),$$

with $C = \{c_\mathbf{j}\} \in \ell^p$, to the cardinal interpolation problem $s(\mathbf{j}) = f(\mathbf{j})$, $\mathbf{j} \in \mathbf{Z}^s$, can be formulated as

$$(9.11) \qquad s(\mathbf{x}) = \sum_{\mathbf{j} \in \mathbf{Z}^s} f(\mathbf{j}) L(\mathbf{x} - \mathbf{j})$$

where the series converges uniformly on compact subsets of $\mathbf{R}^s$.

Remark. Since $L(\mathbf{x})$ has exponential decay, the conclusion that the series (9.10) converges uniformly on compact subsets of $\mathbf{R}^s$ to $s(\mathbf{x}) \in S(\phi)$ which satisfies $s(\mathbf{j}) = f(\mathbf{j})$ for all $\mathbf{j} \in \mathbf{Z}^s$ still holds, even when the data sequence $\{f(\mathbf{j})\}$ has polynomial growth.

To compute the solution to the cardinal interpolation problem efficiently, we return to the Neumann series approach discussed in §8.4; namely, by setting

$$M = I - \Phi = \{\delta_{\mathbf{j}0} - \phi(\mathbf{j})\},$$

the coefficient sequence $C = \{c_\mathbf{j}\}$ in (9.10) that provides the solution to the cardinal interpolation problem $s(\mathbf{j}) = f(\mathbf{j})$, satisfies, at least formally:

$$(9.12) \qquad C = (I + M + \cdots) F$$

where $F = \{f(\mathbf{j})\}$. That the series (9.12) actually converges to C follows from the assumption (i) - (iii) on ϕ. Indeed, by (i) and (ii), we have

$$|\tilde{\phi}(w)| = \left| \sum_{\mathbf{j} \in \mathbf{Z}^s} \phi(\mathbf{j}) e^{-i\mathbf{j} \cdot \omega} \right|$$
$$\leq \sum_{\mathbf{j} \in \mathbf{Z}^s} |\phi(\mathbf{j})| = \sum_{\mathbf{j} \in \mathbf{Z}^s} \phi(\mathbf{j}) = 1,$$

so that, in view of (iii), it follows that for all $\mathbf{j} \in \mathbf{Z}^s$,

$$(9.13) \qquad |M^k(\mathbf{j})| = \frac{1}{(2\pi)^s} \left| \int_{(-\pi,\pi)^s} \widetilde{M}^k(\omega) e^{i\mathbf{j} \cdot \omega} dw \right|$$
$$= \frac{1}{(2\pi)^s} \left| \int_{(-\pi,\pi)^s} (1 - \tilde{\phi}(w))^k e^{i\mathbf{j} \cdot \omega} dw \right|$$
$$\leq r^k \longrightarrow 0$$

where

$$r = 1 - \min_{\omega \in \mathbf{R}^s} \tilde{\phi}(\omega)$$

satisfies $0 < r < 1$. The estimate in (9.13) also helps us decide how large k must be chosen so that the kth partial sum of the series in (9.12), namely:

$$C_k = (I + M + \cdots + M^k)F$$
$$= \wedge_k F,$$

would already give a satisfactory estimate of C. By this we mean, of course, that the "quasi-interpolation" operator

$$(P_k f)(\mathbf{x}) = \sum_{\mathbf{j} \in \mathbf{Z}^s} (\wedge_k F)(\mathbf{j}) \phi(\mathbf{x} - \mathbf{j})$$
$$= \sum_{\mathbf{j} \in \mathbf{Z}^s} C_k(\mathbf{j}) \phi(\mathbf{x} - \mathbf{j})$$

introduced in §8.4 would provide a satisfactory approximation to the cardinal interpolant

$$(P_\infty f)(\mathbf{x}) = \sum_{\mathbf{j} \in \mathbf{Z}^s} C(\mathbf{j}) \phi(\mathbf{x} - \mathbf{j})$$

at the nodes $\mathbf{x} = \mathbf{m} \in \mathbf{Z}^s$. Indeed, suppose that the commutator of ϕ has degree at least $n_0 \in \mathbf{Z}_+$. Then for $k \geq \frac{1}{2}(n_0 - 1)$ (see Theorems 8.7 and 8.8 for symmetric ϕ) and any polynomial $p \in \pi_{n_0}^s$, we have:

(9.14)
$$|(P_k f - P_\infty f)(\mathbf{m})|$$
$$= |(P_k f - f)(\mathbf{m})|$$
$$= |\{P_k(f - p) - (f - p)\}(\mathbf{m})|$$
$$= |\{(I + \cdots + M^k)(F - P)\Phi - (F - P)\}(\mathbf{m})|$$
$$= |\{(I + \cdots + M^k)(I - M) - I\}(F - P)(\mathbf{m})|$$
$$= |M^{k+1}(F - P)(\mathbf{m})|$$
$$\leq r^{k+1} \operatorname{dist}_{\ell^\infty}(f, \pi_{n_0}^s),$$

where $\mathbf{m} \in \mathbf{Z}^s$ and we have used the notation $P = \{p(\mathbf{j})\}$. Of course, the ℓ^∞ distance is taken at $\mathbf{Z}^s$. Hence, even if we are modest by choosing $n_0 = 0$ in (9.14), we still have

(9.15)
$$\left| \sum_{\mathbf{j} \in \mathbf{Z}^s} C_k(\mathbf{j}) \phi(\mathbf{m} - \mathbf{j}) - f(\mathbf{m}) \right|$$
$$\leq r^{k+1} \sup_{\mathbf{j} \in \mathbf{Z}^s} |f(\mathbf{j}) - \overline{f}|$$

for all $\mathbf{m} \in \mathbf{Z}^s$, where

$$\overline{f} = \frac{1}{2}(\sup_{\mathbf{m}\in\mathbf{Z}^s} f(\mathbf{m}) + \inf_{\mathbf{m}\in\mathbf{Z}^s} f(\mathbf{m})).$$

Consequently, if the upper and lower bounds of the data sequence F do not differ by too much, it does not take a very large value of k for the kth quasi-interpolant $P_k f$ to give a good approximation of the cardinal interpolant $P_\infty f$ at the nodes $\mathbf{Z}^s$.

For more details and suggestions on numerical implementation, the reader is referred to Chui, Diamond, and Raphael [54] and the appendix to this monograph.

9.4. Cardinal interpolation with singular ϕ. In this section we will study the situation where ϕ is singular, or more precisely, the zero set of the symbol σ_ϕ of ϕ has nonempty intersection with T^s. Of course, one does not expect to have unique cardinal interpolants in general. Our approach is to give a stable computational scheme of a cardinal interpolant that reproduces all polynomials in $\pi_{n_o}^s$ where n_o is the commutator degree of ϕ. Hence, the maximum order of approximation is guaranteed by scaled cardinal interpolation, a topic to be discussed in the next section. Since partial sums of the Neumann series already reproduce $\pi_{n_o}^s$, we will study convergence of this sequence, even in the situation where ϕ is singular; and in case of divergence, we will have to modify this sequence either by subtracting from it a (null) sequence (i.e., one that produces an interpolant of the zero data) or by adjusting the negative part of the discrete Fourier transform of each term so that the new sequence converges to the required cardinal interpolant. The results in this section are essentially contained in Chui, Diamond, and Raphael [54]. To facilitate our estimates, we will make the following assumptions on ϕ:

(i) $\sum_{\mathbf{j}\in\mathbf{Z}^s} \phi(\mathbf{j}) = 1$,

(ii) $\phi(\mathbf{j}) \geq 0$ and $\phi(-\mathbf{j}) = \phi(\mathbf{j})$ for all $\mathbf{j} \in \mathbf{Z}^s$,

(iii) The commutator of ϕ has degree $n_0 \in \mathbf{Z}_+$.

It should be noted that the symmetry assumption is not restrictive since the average

$$\frac{1}{2}(\phi(\mathbf{x}) + \phi(-\mathbf{x}))$$

is always symmetric and preserves the other properties in (i) - (iii).

In the following, recall that $M = I - \Phi$.

LEMMA 9.3. *Let $\tilde{\phi}(w) \geq 0$ and have only isolated zeros in $[-\pi,\pi]^s$, such that the Hessian $H_{\mathbf{j}}$ of $\tilde{\phi}(w)$ at each zero $w_{\mathbf{j}}$ is positive definite. Then*

$$(9.16) \quad M^n(\mathbf{k}) = \frac{1}{(2\pi n)^{s/2}} \sum_{\mathbf{j}} \frac{e^{-i\mathbf{k}\cdot w_{\mathbf{j}}}}{|H_{\mathbf{j}}|^{1/2}} e^{-\mathbf{k}^T H_{\mathbf{j}}^{-1} \mathbf{k}/2n}$$

$$\times \left\{ 1 + \sum_{|\alpha|=3}^{m} P_{\alpha,\mathbf{j}}(n) Q_{\alpha,\mathbf{j}}(\frac{1}{\sqrt{n}}\mathbf{k}) n^{-|\alpha|/2} \right\}$$

$$+ O(n^{[(m+1)/3]-(m+1+s)/2})$$

where $P_{\alpha,\mathbf{j}} \in \pi_{[|\alpha|/3]}^s, Q_{\alpha,\mathbf{j}} \in \pi_{|\alpha|}^s$, $[x]$ denotes the integer part of x, and the approximation in (9.16) *is uniform in* $\mathbf{k}$.

Here, m is arbitrary but fixed. We will only sketch the idea of its proof. To estimate $M^n(\mathbf{k})$, we use the ideas of Laplace's method in asymptotic analysis to estimate

$$\int_{(-\pi,\pi)^s} \widetilde{M}^n(w) e^{-i\omega\cdot\mathbf{k}} d\omega.$$

Note that since $\widetilde{M}(\omega)$ attains its maximum value, which is one, at precisely the isolated zeros $w_{\mathbf{j}}$ of $\tilde{\phi}$, we have $\widetilde{M}^n(w) \to 0$ exponentially outside any fixed neighborhood of $\omega_{\mathbf{j}}$. Hence, we only have to consider

$$\int_{|\omega|<\varepsilon} \widetilde{M}^n(\omega + \omega_{\mathbf{j}}) e^{-i\mathbf{k}\cdot(\omega+\omega_{\mathbf{j}})} d\omega.$$

By choosing $\varepsilon = n^{-2/5}$, we have:

(a) $\widetilde{M}^n(\omega + \omega_{\mathbf{j}}) < (1 - c|\omega|^2)^n < e^{-cn^{1/5}} \to 0$ for $|\omega| \geq \varepsilon$,
(b) $n\varepsilon^2 = n^{1/5} \to \infty$, and
(c) $n\varepsilon^3 = n^{-1/5} \to 0$.

Hence, using Taylor's expansion in $|\omega| < \varepsilon = n^{-2/5}$, we have

$$(9.17) \quad \widetilde{M}^n(\omega + \omega_{\mathbf{j}})$$

$$= \exp\left\{ n \log\left(1 - \frac{\omega^T H_{\mathbf{j}} \omega}{2} + \sum_{|\alpha|=3}^{m} \frac{D^\alpha \widetilde{M}(\omega_{\mathbf{j}})}{\alpha!} \omega^\alpha + O(|\omega|^{m+1}) \right) \right\}$$

$$= \exp\left(-\frac{n\omega^T H_{\mathbf{j}} \omega}{2} \right) \left\{ 1 + \sum_{|\alpha|=3}^{m} P_\alpha(n) \omega^\alpha + O\left(n^{[\frac{m+1}{3}]} |\omega|^{m+1} \right) \right\}.$$

Now multiplying both sides by $e^{-i\mathbf{k}\cdot(\omega+\omega_\mathbf{j})}$, integrating over $|\omega|<\epsilon$, and extending the integral to all of $\mathbf{R}^s$ by introducing an error which is exponentially small in n, we are led to the consideration of integrals of the form

$$\int_{\mathbf{R}^s} e^{-n\frac{\omega^T H_\mathbf{j}\omega}{2}} e^{-i\mathbf{k}\cdot\omega}\omega^\alpha d\omega$$

which can be integrated to be

$$\left(\frac{2\pi}{n}\right)^{s/2} \frac{1}{|H_\mathbf{j}|^{1/2}} \exp\{-\frac{1}{2}\mathbf{k}^T H_\mathbf{j}^{-1}\mathbf{k}\} \frac{1}{n^{|\alpha|/2}} Q_{\alpha,\mathbf{j}}(k/\sqrt{n})$$

where $Q_{\alpha,\mathbf{j}} \in \pi_{|\alpha|}^s$. The error term $O(|\omega|^{m+1})$ in (9.17) gives a contribution of $O(n^{-\frac{m+1+s}{2}})$ when it is integrated against $e^{-i\mathbf{k}\cdot\omega}$. This completes a sketch of the proof of Lemma 9.3.

As a consequence of this lemma, we note the following:

Case (1). If $s > 2$, then $M^n(\mathbf{k}) \to 0$ uniformly in $\mathbf{k}$ and the Neumann series

(9.18) $$I + M + \cdots$$

converges uniformly to $\wedge^*$ that satisfies

(9.19) $$\wedge^* \Phi = I.$$

In the following, if $\wedge^*$ satisfies (9.18) it will be called a *fundamental sequence*.

Case (2). If $s = 2$, then the Neumann series

$$I + M + \cdots$$

diverges. Fortunately, it also follows from Lemma 9.3 that the series

(9.20) $$\sum_{n=0}^{\infty} \left\{ M^n(\mathbf{k}) - \frac{1}{2\pi(n+1)} \sum_{\mathbf{j}} |H_\mathbf{j}|^{-\frac{1}{2}} e^{-i\mathbf{k}\cdot\omega_\mathbf{j}} \right\}$$

converges uniformly in $\mathbf{k}$ to a fundamental sequence $\wedge^*$.

By using the Euler–Maclaurin summation formula, we can also determine the asymptotic behaviors of $\wedge^*$.

THEOREM 9.9. *Let the hypothesis in Lemma 9.3 be satisfied. Then the fundamental sequences $\wedge^*$ defined as the uniform limit of the Neumann series (9.18) for $s > 2$ and as the uniform limit of the series (9.20) for $s = 2$ have the following asymptotic properties:*

Case (1). For $s > 2$,

$$\wedge^*(\mathbf{k}) = \frac{1}{(2\pi)^{s/2}} \sum_{\mathbf{j}} \frac{e^{-i\mathbf{k}\cdot\omega_{\mathbf{j}}}}{|H_{\mathbf{j}}|^{1/2}} \{\frac{1}{2}\mathbf{k}^T H_{\mathbf{j}}^{-1}\mathbf{k}\}^{1-\frac{s}{2}} \tag{9.21}$$

$$\times \Gamma(\frac{s}{2} - 1)(1 + O(\frac{1}{|\mathbf{k}|}))$$

as $|\mathbf{k}| \to \infty$.

Case (2). For $s = 2$,

$$\wedge^*(\mathbf{k}) = \frac{-1}{2\pi} \sum_{\mathbf{j}} \frac{e^{-i\mathbf{k}\cdot\omega_{\mathbf{j}}}}{|H_{\mathbf{j}}|^{1/2}} \left\{ 2\gamma + \log\left(\frac{\mathbf{k}^T H_{\mathbf{j}}^{-1}\mathbf{k}}{2}\right) \right\} + O(\frac{1}{|\mathbf{k}|}) \tag{9.22}$$

as $|\mathbf{k}| \to \infty$, where γ is Euler's constant.

Hence, for dimension s higher than 2, the sequence $\wedge^*$ actually decays, and in $\mathbf{R}^2$, $\wedge^*$ only has logarithmic growth. This information is very important in characterizing what data sequence $F = \{f(\mathbf{j})\}$ admits bounded, or more generally $L^p(\mathbf{R}^s)$, cardinal interpolants of the form

$$\sum_{\mathbf{j}\in\mathbf{Z}^s} (\wedge^* F)(\mathbf{j})\phi(\mathbf{x} - \mathbf{j}). \tag{9.23}$$

Instead of going into the details in this direction, we will delay it to our discussion on scaled cardinal interpolation.

If the function ϕ is very much singular in the sense that its discrete Fourier transform $\tilde{\phi}(\omega)$ takes on both positive and negative values, then the Neumann series diverges very badly. In this situation it requires more than a modification by an additive null sequence as in Case (2) of Theorem 9.9. The following result is also contained in Chui, Diamond, and Raphael [54].

THEOREM 9.10. *Suppose that the zero set $\{\omega: \tilde{\phi}(\omega) = 0\}$ is a manifold of codimension one on which the gradient of $\tilde{\phi}$ does not vanish. Then the series*

$$\sum_{n=0}^{\infty} \frac{1}{2\pi^s} \left\{ \int_{\{\tilde{\phi}(\omega)>0\}} \widetilde{M}^n(\omega) e^{-i\mathbf{k}\cdot\omega} d\omega \right. \tag{9.24}$$

$$\left. - \int_{\{\tilde{\phi}(\omega)<0\}} (2 - \widetilde{M}(\omega))^n e^{-i\mathbf{k}\cdot\omega} d\omega \right\}$$

converges uniformly in $\mathbf{k} \in \mathbf{Z}^s$ *to a fundamental sequence* $\wedge^*$ *that satisfies* $\wedge^*\Phi = I$. *Furthermore,* $\wedge^*(\mathbf{k}) = O(|\mathbf{k}|)$ *as* $|\mathbf{k}| \to \infty$.

Of course the domains of integration in (9.24) are subsets of $(-\pi,\pi)^s$. Note that the asymptotic property of $\wedge^*$ is not sharp but is sufficient in determining what sequences admit $L^\infty(\mathbf{R}^s)$ cardinal interpolants. Further research is required when the assumption on the zero set of $\tilde{\phi}$ is more relaxed.

Recently, there has been much interest in various aspects of multivariate cardinal interpolation from $S(\phi)$. However, there does not seem to be any result on interpolation from $S(\phi_1,\cdots,\phi_m)$ where $\phi_1,\cdots,\phi_m$ are compactly supported continuous functions in $\mathbf{R}^s$, such as all the box splines or minimal and quasi-minimal supported splines in the same space. This important area of research deserves some attention. We mention the following aspects of cardinal interpolation without going into any details:

(1) de Boor, Höllig, and Riemenschneider [37]–[39] and Höllig, Marsden, and Riemenschneider [134] study convergence of cardinal interpolants from $S(\phi)$, where ϕ is a box spline, as the degree of the box spline tends to infinity.

(2) Riemenschneider and Scherer [170] study cardinal Hermite interpolation with box splines, extending the univariate considerations by Schoenberg [180], Schoenberg and Sharma [181], and Lee [144].

(3) Jetter and Riemenschneider [139], [140] study the cardinal interpolation problem on submodules of $\mathbf{Z}^s$, instead of all of $\mathbf{Z}^s$.

(4) Sivakumar [192] and Stöckler (manuscript in preparation) consider the cardinal interpolation by shifted box splines on the three-directional mesh. Stöckler's result is stated in Jetter [138].

(5) ter Morsche [161] investigates cardinal interpolation of periodic data, and in particular, relates the Fourier coefficients of the error term in the cardinal interpolation with those of the commutator. Consequently, by scaling the interpolants, an asymptotic expression of these Fourier coefficients is obtained.

9.5. Scaled cardinal interpolation. Recently scaled cardinal interpolation, or interpolation from

$$S_h(\phi) = \{\sigma_h f : f \in S(\phi)\}, \quad h > 0,$$

at the nodes $h\mathbf{j}$, $\mathbf{j} \in \mathbf{Z}^s$, has also been investigated. (See, for example, Chui, Jetter, and Ward [63] and ter Morsche [160].) Since we are interested in the computational aspect in both the singular and nonsingular cases, we will only consider interpolants by scaling the cardinal interpolants discussed in §8.3 and §8.4. The objective is the following. Let W be a closed set in $\mathbf{R}^s$ and f a sufficiently smooth function on some open set Ω where $W \subset \Omega$. We would like to obtain an $s_h \in S_h(\phi)$ that satisfies:

(9.25)
$$\begin{cases} (s_h - f)(\mathbf{j}h) = 0, \quad h\mathbf{j} \in W, \\ \text{and} \\ \| s_h - f \|_K = O(h^{n_0+1}), \\ \text{for any compact set } K \subset \widetilde{W}. \end{cases}$$

Here, n_0 denotes the commutator degree of ϕ, so that the order of approximation from $S_h(\phi)$ is required to be optimal.

We will follow the treatment in Chui, Diamond, and Raphael [54]. To facilitate our presentation, we need the following notation. Let $\rho \in \mathbf{Z}_+$ and set

$$C^{\rho,1}(\Omega) = \{g \in C^\rho(\Omega): \ D^\alpha g \in \text{Lip}_\Omega(1), \ |\alpha| = \rho\}.$$

Define $F_h(\mathbf{k})$ by

$$F_h(\mathbf{k}) = \begin{cases} f(h\mathbf{k}) & \text{for } h\mathbf{k} \in \Omega \\ 0 & \text{otherwise}. \end{cases}$$

In addition, let Λ^* denote a fundamental sequence defined by (9.18), (9.20), or (9.24) depending on the appropriate situations. In the nonsingular case, we can establish the following result:

THEOREM 9.11. *Let Ω be open and $f \in C^{n_0,1}(\Omega)$ be polynomially bounded. If $\tilde{\phi}(\omega) > 0$, then the scaled cardinal interpolant*

$$s_h(\mathbf{x}) = \sum_{\mathbf{j} \in \mathbf{Z}^s} (\Lambda^* F_h)(\mathbf{j}) \phi(\mathbf{x}/h - \mathbf{j})$$

of F_h satisfies $\| s_h - f \|_K = O(h^{n_0+1})$ for any compact set $K \subset \Omega$.

To demonstrate this result, note first that for any n, $\Lambda^* F_h = \Lambda_n F_h + \Lambda^* M^{n+1} F_h$, where $\Lambda_n = I + \cdots + M^n$. If $n \geq (n_0 - 1)/2$, then $\Lambda_n F_h$ gives the coefficients of the quasi-interpolant $P_n f$ and $M^{n+1} F_h = O(h^{n_0+1})$ in the interior of Ω. Since Λ^* decays exponentially, $\Lambda^* M^{n+1} F_h(\mathbf{j}) = O(h^{n_0+1})$ for $h\mathbf{j}$ in some open neighborhood of K away from the boundary of Ω. Finally, for $n \geq (n_0 - 1)/2$, we have

$$\begin{aligned}
\| s_h - f \|_K &\leq \| f - \sum_{\mathbf{j} \in \mathbf{Z}^s} (\Lambda_n F_h)(\mathbf{j}) \phi(\cdot/h - \mathbf{j}) \|_K \\
&+ \| \sum_{\mathbf{j} \in \mathbf{Z}^s} (\Lambda^* M^{n+1} F_h)(\mathbf{j}) \phi(\cdot/h - \mathbf{j}) \|_K \\
&= O(h^{n_0+1}).
\end{aligned}$$

The singular cases require a bit more care when considering interpolation on W because the fundamental sequences do not decay exponentially. In these cases we construct our interpolant as the sum of the quasi-interpolant $(P_n f)(\mathbf{x})$ plus an interpolant of the error

$$E_n(\mathbf{k}) = f(h\mathbf{k}) - (P_n f)(h\mathbf{k}) = M^{n+1} F_h(\mathbf{k})$$

at the grid nodes $h\mathbf{k} \in W$. Leaving for later considerations of convergence, we define the interpolants $s_h(\mathbf{x})$ as follows:

Case (1). $s > 2, \tilde{\phi}(\omega) \geq 0$.

(9.26)
$$\begin{aligned}
s_h(\mathbf{x}) &= (P_n f)(\mathbf{x}) \\
&+ \sum_{\mathbf{j} \in \mathbf{Z}^s} \left(\sum_{h\mathbf{k} \in W} \Lambda^*(\mathbf{j} - \mathbf{k}) E_n(\mathbf{k}) \right) \phi(\mathbf{x}/h - \mathbf{j}),
\end{aligned}$$

for $n \geq (n_0 + 1)/2$.

Case (2). $s = 2$, $\tilde{\phi}(\omega) \geq 0$.

We first define a modified fundamental sequence $\Lambda^*(\mathbf{j}; \mathbf{k})$, where $\mathbf{k}$ is a parameter, by

(9.27)
$$\begin{aligned}
\Lambda^*(\mathbf{j}; \mathbf{k}) &= \log(1 + |\mathbf{k}|^2) \sum_\ell e^{-i\mathbf{k}\cdot\omega_\ell} \frac{1}{2\pi |H_\ell|^{1/2}} \\
&+ \Lambda^*(\mathbf{j} - \mathbf{k})
\end{aligned}$$

so that

$$\phi \Lambda^*(\mathbf{j}; \mathbf{k}) = \begin{cases} 1 & \text{if } \mathbf{j} = \mathbf{k} \\ 0 & \text{otherwise} \end{cases}.$$

The interpolant on W is then defined by

(9.28)
$$s_h(\mathbf{x}) = (P_n f)(\mathbf{x}) + \sum_{\mathbf{j} \in \mathbf{Z}^s} \left(\sum_{h\mathbf{k} \in W} \Lambda^*(\mathbf{j}; \mathbf{k}) E_n(\mathbf{k}) \right) \phi(\mathbf{x}/h - \mathbf{j}),$$

$n \geq (n_0 + 1)/2$.

Case (3). $\tilde{\phi}(\omega)$ vanishes on a manifold of codimension 1.

In this case the interpolant is given by (9.26) without any changes. We have the following result:

THEOREM 9.12. *Let W be a closed subset of the open set $\Omega \subset \mathbf{R}^s$. Then the scaled cardinal interpolants s_h of a function $f \in C(\Omega)$ defined by (9.26), (9.28), (9.29) satisfy*

$$\| s_h - f \|_K = O(h^{n_0+1})$$

where K is a compact set in Ω, provided that f satisfies the following:

Cases (1) *and* (2):
  $f \in C^{n_0+3}(\Omega)$ and $|\mathbf{x}|^{2-s} D^\alpha f(\mathbf{x}) \in L^1(\Omega)$ for $|\alpha| = n_0 + 3$.

Case (3):
  $f \in C^{n_0+s+2}(\Omega)$ and $|\mathbf{x}| D^\alpha f(\mathbf{x}) \in L^1(\Omega)$ for $|\alpha| = n_0 + s + 2$.

The proof of these results consists of showing that the inner sums in (9.26), (9.28), and (9.29) are each $O(h^{n_0+1})$ for bounded $|h\mathbf{j}|$ under the respective hypotheses on f. We use the fact that for $f \in C^m(\Omega)$, $m \leq 2n+2$, there exists a constant c_m independent of $\mathbf{k}$, $h\mathbf{k} \in W$, such that $|E_n(\mathbf{k})| \leq c_m h^m |D^\alpha f(\mathbf{x_k})|$ where $|\alpha| = m$, and $|\mathbf{x_k} - h\mathbf{k}| \leq c_m h$. In Case (1) of Theorem 9.12, using the asymptotic behavior (9.21), we are led to consider the sum

$$\sum_{\substack{h\mathbf{k} \in W \\ \mathbf{k} \neq \mathbf{j}}} |(\mathbf{j} - \mathbf{k})|^{2-s} h^{n_0+3} |D^\alpha f(\mathbf{x_k})|$$

$$= h^{n_0+3-2} \sum_{\substack{h\mathbf{k} \in W \\ \mathbf{k} \neq \mathbf{j}}} |(h\mathbf{j} - h\mathbf{k})|^{2-s} |D^\alpha f(\mathbf{x_k})| h^s = O(h^{n_0+1})$$

where $|\alpha| = n_0 + 3$, using the integrability condition on $|D^\alpha f|$. In Case (2) of this theorem, using the asymptotic behavior (9.22) and the definition

(9.27) of $\Lambda^*(\mathbf{j};\mathbf{k})$, we can write $\Lambda^*(\mathbf{j};\mathbf{k}) = O(\log(|\mathbf{y}|/|\mathbf{x}-\mathbf{y}|))$ where $\mathbf{x} = h\mathbf{j}$ and $\mathbf{y} = h\mathbf{k}$, uniformly in $\mathbf{x}, \mathbf{y}$. We can then estimate

$$\sum_{h\mathbf{k}\in W} \Lambda^*(\mathbf{j};\mathbf{k}) E_n(\mathbf{k})$$
$$= h^{n_0+3-2} \sum_{\substack{h\mathbf{k}\in W \\ \mathbf{k}\neq 0,\mathbf{j}}} O(\log|\mathbf{y}|/|\mathbf{x}-\mathbf{y}|)|D^\alpha f(\mathbf{x_k})|h^2 = O(h^{n_0+1})$$

using the integrability condition. Finally, in Case (3) of Theorem 9.12, we consider the sum

$$h^{n_0+s+2} \sum_{h\mathbf{k}\in W} O(|\mathbf{j}-\mathbf{k}|)|D^\alpha f(\mathbf{x_k})|$$
$$= h^{n_0+1} \sum_{h\mathbf{k}\in W} O(|\mathbf{x}-\mathbf{y}|)|D^\alpha f(\mathbf{x_k})|h^s = O(h^{n_0+1})$$

as required.

If W is bounded, the requirements on f in Theorem 9.12 can be relaxed to $f \in C^{n_0+2,1}(\Omega)$ in Cases (1) and (2) and $f \in C^{n_0+s+1,1}(\Omega)$ in Case (3) (see [54] for further details).

CHAPTER 10

Shape-Preserving Approximation and Other Applications

Spline functions in one variable have found numerous applications in virtually all areas of sciences and technologies, namely: curve fitting, finite element methods (FEM), numerical solutions of differential and integral equations, computer aided geometric design (CAGD), signal processing, mathematical modeling, etc. We have already discussed approximation and interpolation of both continuous and discrete data by multivariate splines in Chapters 6, 8, and 9. In particular, vertex splines and generalized vertex splines can be applied to approximation of scattered data, such as least-squares approximation, that requires a spline series of locally supported elements. Another application of a spline series is to reconstruct a gradient field. This problem will be introduced later in this chapter. The approximating "surfaces" so obtained can be displayed by computing Bézier nets on dyadic subdivision of the simplices they are defined on by using the method described in §7.1. If the approximants or interpolants happen to be some box spline series, the "surfaces" can be displayed very efficiently by applying a subdivision algorithm such as the line average algorithm discussed in §7.3. Multivariate splines, and in particular vertex splines, can be used to construct smooth elements, that satisfy certain constraints on the derivatives. Due to the attractive structure of these elements such as local supportedness and Bézier representation, there are important applications to FEM and CAGD. In CAGD, for instance, it is required to construct "surfaces" that satisfy certain constraints or shape characteristics. If the prescribed shape characteristics are described by using a finite number of data values, then the problem of shape-preserving approximation or interpolation has to be solved. There has been much interest in this problem recently. On the other hand, however, there does not seem to be any serious attempt in applying multivariate splines, other than the tensor-product splines, in numerical solution of partial differential equations, signal processing, and mathematical modeling.

10.1. Shape-preserving approximation by box spline series.
We first recall from §1.4 that the univariate "variation diminishing" spline

approximant Vf of any data function f follows the "shape" of f in the sense that

$$f \geq 0 \implies Vf \geq 0,$$
$$f \uparrow \implies Vf \uparrow,$$

and

$$f \text{ convex} \implies Vf \text{ convex}.$$

In the special situation when the knot sequence is equally spaced, Vf is defined as a B-spline series with coefficients given by the values of the data function f at the centers of the supports of the B-splines. Let us study the analogous situation in the bivariate setting from the following simple example.

Example 10.1. Consider the bivariate spline space $S_1^0(\Delta^{(1)})$ of piecewise linear polynomials on the three-directional mesh $\Delta^{(1)}$ with grid points at $\mathbf{Z}^2$. We have seen in §3.2 that a basis of this space is given by $M_{111}(x-i, y-j)$, $i,j \in \mathbf{Z}$, where $M_{111}(x,y)$ is the Courant hat function with center at the origin. Let $f(x,y)$ be a data function and consider the bivariate spline approximant

$$(Vf)(x,y) = \sum_{i,j} f(i,j) M_{111}(x-i, y-j).$$

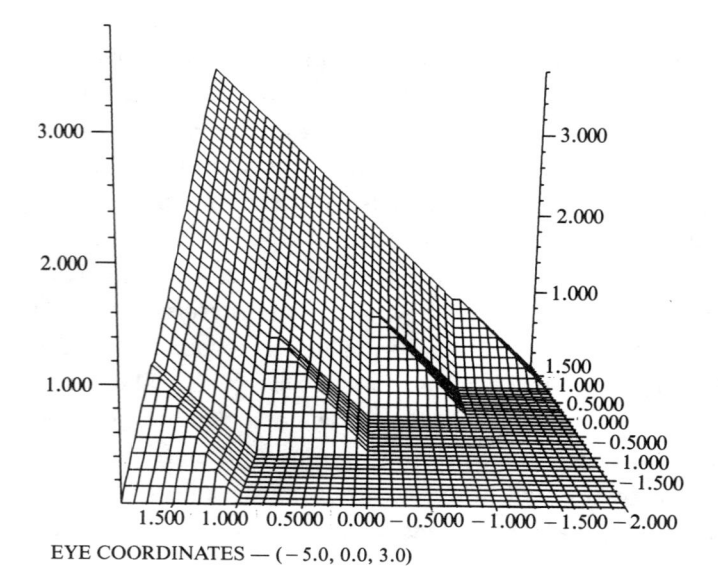

FIG. 10.1

Hence, Vf is the bivariate analogue of Schoenberg's univariate variation diminishing spline. Let us consider the data function

$$f(x,y) = (x+y)_+ ,$$

which is certainly monotone in any direction and is also convex. However, its approximant $V(f)$ zigzags in the region

$$\{ (x,y): \ -1 < x+y < 0 \},$$

(cf. Dahmen and Micchelli [100], and see Figure 10.1).

From the above simple example, we see that the problem of constructing a shape-preserving approximation scheme in the multivariate setting needs more careful investigation. There are at least two lessons to be learned from this example. First, the information on the discrete data alone does not necessarily tell the shape characteristics of the data function. It also takes a triangulation (or in the multivariate case, a simplicial partition) with vertices at the sample points (or nodes) to give us the precise information. In the above example when the discrete data is

$$\{(i+j)_+ : \ i,j \in \mathbf{Z}\},$$

if we use the three-directional mesh given by the direction set $\{\mathbf{e}^1, \mathbf{e}^2, \mathbf{e}^1 + \mathbf{e}^2\}$, then the data set yields neither a monotone nor convex interpolant by piecewise linear polynomials. However, if we choose the direction set to be $\{\mathbf{e}^1, \mathbf{e}^2, \mathbf{e}^2 - \mathbf{e}^1\}$, then the piecewise linear polynomial interpolant at the vertices of this mesh is precisely $(x+y)_+$, which is both monotone in any direction and convex. Second, the direction set that determines the shape of the discrete data should be used at least as a subgrid of the grid partition for the multivariate spline approximant.

DEFINITION. A discrete data set $\{(\mathbf{x}^i, f_i)\}$ is said to be monotone in a given direction ξ (and/or convex) if there exists a simplicial partition Δ satisfying (i) and (ii) in §6.1 whose vertices consist only of $\{\mathbf{x}^i\}$ such that the piecewise linear polynomial on this partition interpolating f_i at $\mathbf{x}^i$ is monotone in the direction ξ (and/or convex, respectively). If it happens that Δ can be chosen to be an $(s+1)$ directional mesh with direction set X_{s+1}, we will say that the data set is *regularly monotone* in the direction ξ (and/or *regularly convex*) with direction set X_{s+1}.

The data set $\{(i,j), \ (i+j)_+\}$ in Example 10.1 is both regularly monotone in any direction and regularly convex, with direction set $X_3 = \{\mathbf{e}^1, \mathbf{e}^2, \mathbf{e}^2 - \mathbf{e}^1\}$.

DEFINITION. Let $X_{s+1} \subset \mathbf{Z}^s \backslash \{\mathbf{0}\}$ with $\langle X_{s+1} \backslash \mathbf{y} \rangle = \mathbf{R}^s$ for all $\mathbf{y} \in X_{s+1}$. Then the box spline series

$$s_0(\mathbf{x}) = \sum_{\mathbf{i} \in \mathbf{Z}^s} c_\mathbf{i} M(\mathbf{x} - \mathbf{i} | X_{s+1}) \tag{10.1}$$

is called a *control polygon*.

Note that $s_0(\mathbf{x})$ is a piecewise linear polynomial and hence defines a certain shape. The following result due to Dahmen and Micchelli [100] clarifies this terminology.

THEOREM 10.1. *Let $X_{s+1} \subset \mathbf{Z}^s \backslash \{\mathbf{0}\}$ with $\langle X_{s+1} \backslash \mathbf{y} \rangle = \mathbf{R}^s$ for any $\mathbf{y} \in X_{s+1}$. Then for any direction set $X_n \subset \mathbf{Z}^s \backslash \{\mathbf{0}\}$ with $X_n \supseteq X_{s+1}$, the box spline series*

$$s(\mathbf{x}) = \sum_{\mathbf{i} \in \mathbf{Z}^s} c_\mathbf{i} M(\mathbf{x} - \mathbf{i} | X_n) \tag{10.2}$$

preserves positivity, monotonicity, and convexity of the control polygon defined in (10.1).

The proof of this result is quite easy. In fact, by defining

$$s_1(\mathbf{x}) = \sum_{\mathbf{i} \in \mathbf{Z}^s} c_\mathbf{i} M(\mathbf{x} - \mathbf{i} | X_{s+1} \cup \{\mathbf{x}^*\}),$$

we have

$$s_1(\mathbf{x}) = \int_{-\frac{1}{2}}^{\frac{1}{2}} s_0(\mathbf{x} - t\mathbf{x}^*) dt$$

which certainly preserves positivity, monotonicity, and convexity of the control polygon $s_0(\mathbf{x})$. Consequently, by adding directions one at a time to X_{s+1}, $s(\mathbf{x})$ defined in (10.2) must preserve the same shape characteristics of $s_0(\mathbf{x})$.

Hence, it is important to start with a control polygon with the desirable shape before we add directions to give a smoother multivariate spline approximant. In the following we give some sufficient conditions on the data sequence that guarantee the monotone or convex shapes of the control polygon. For convenience, we only consider

$$X_{s+1} = \{\mathbf{e}^1, \cdots, \mathbf{e}^s, \mathbf{e}^1 + \cdots + \mathbf{e}^s\} \tag{10.3}$$

where $\{\mathbf{e}^i\}$ is the standard basis of $\mathbf{R}^s$. Of course, X_3 gives the three-directional mesh in $\mathbf{R}^2$. The following result is also in Dahmen and Micchelli [100].

THEOREM 10.2. *Let X_{s+1} be as given in (10.3). Then the following conclusions can be made.*

(i) *For any $\xi \in X_{s+1}$, if $c_{\mathbf{i}+\xi} \geq c_{\mathbf{i}}$ for all $\mathbf{i} \in \mathbf{Z}^s$, then the control polygon s_0 is monotone nondecreasing in the direction ξ.*

(ii) *If $\{c_{\mathbf{i}}\}$ satisfies*

$$(10.4) \qquad c_{\mathbf{i}} + c_{\mathbf{i}+\mathbf{e}^\ell - \mathbf{e}^m} \geq c_{\mathbf{i}+\mathbf{e}^\ell} + c_{\mathbf{i}-\mathbf{e}^m}$$

for all $\mathbf{i} \in \mathbf{Z}^s$ and all $\mathbf{e}^\ell, \mathbf{e}^m$ with $l \neq m$ and $\mathbf{e}^{s+1} = -(\mathbf{e}^1 + \cdots + \mathbf{e}^s)$, then the control polygon is convex.

The proof of (i) is trivial since by Theorem 2.4, we have

$$\begin{aligned}
D_\xi(s_0)(\mathbf{x}) &= \sum_{\mathbf{j} \in \mathbf{Z}^s} c_{\mathbf{j}} D_\xi M(\mathbf{x} - \mathbf{j} | X_{s+1}) \\
&= \sum_{\mathbf{j} \in \mathbf{Z}^s} c_{\mathbf{j}} \Delta_\xi M(\mathbf{x} - \mathbf{j} | X_{s+1} \setminus \{\xi\}) \\
&= \sum_{\mathbf{j} \in \mathbf{Z}^s} c_{\mathbf{j}} \left(M(\mathbf{x} + \frac{1}{2}\xi - \mathbf{j} | X_{s+1} \setminus \{\xi\}) \right. \\
&\qquad \left. - M(\mathbf{x} + \frac{1}{2}\xi - \mathbf{j} - \xi | X_{s+1} \setminus \{\xi\}) \right) \\
&= \sum_{\mathbf{j} \in \mathbf{Z}^s} (c_{\mathbf{j}+\xi} - c_{\mathbf{j}}) M(\mathbf{x} + \frac{1}{2}\xi - \mathbf{j} | X_{s+1} \setminus \{\xi\}) \\
&\geq 0.
\end{aligned}$$

The proof of (ii) is a little more involved and we will omit it here. The reader is referred to Dahmen and Micchelli [100].

10.2. Shape-preserving quasi-interpolation and interpolation. In many applications, the data function f may be strictly positive (or lying on one side of a hyperplane), strictly monotone, and/or strictly convex. Hence, any good spline approximant of f will have the same shape characteristics of f if we allow the grid spacing to be arbitrarily small. For application purposes, however, the mesh size must be known. In this section, we are interested in sharp estimates on the grid spacing h such that the quasi-interpolants $(P_k^h f)(\mathbf{x})$ defined in (8.35) and interpolants studied in §9.3 preserve the same shape characteristics of f. Since these approximants and interpolants also provide the optimal order of approximation, this study should find important applications in CAGD, surface fitting, and building mathematical models. The material in this section is discussed in Chui, Diamond, and Raphael [55]. Because the estimates are very involved, only the three- and four-directional meshes in $\mathbf{R}^2$ are considered, and in order to use the Neumann series, the four-directional mesh is not used in constructing interpolants (see §9.4 and §9.5).

Of course, certain smoothness conditions must be assumed on the data function in order to give a high order of approximation. The notation $C^{\rho,1}$ introduced in §9.5 will be used here. In addition, the grid spacing $h > 0$ must also depend on how strict the positivity, monotonicity, or convexity of the data function f is given; that is, we must assume some knowledge of an $\varepsilon > 0$, such that

(i) $f \geq \varepsilon > 0$;
(ii) $\mathbf{u}_0 \cdot \nabla f \geq \varepsilon > 0$; where $\mathbf{u}_0 = \xi/|\xi|$ is a unit vector that indicates the direction along which f increases; or
(iii) $D_{\mathbf{u}}^2 f \geq \varepsilon > 0$ for all unit vectors $\mathbf{u}$, where $D_{\mathbf{u}}^2 f = \mathbf{u}^T H_f \mathbf{u}$ and

$$H_f = \begin{bmatrix} f_{xx} & f_{xy} \\ f_{xy} & f_{yy} \end{bmatrix}$$

is the Hessian matrix of f, respectively.

Instead of starting with a control polygon as in §10.1, we start with a box spline series using both M_{221} and M_{122} for the three-directional mesh and M_{1111} for the four-directional mesh, since the second derivatives must be taken in studying convexity. By adding directions, we still preserve the same shape as indicated by the proof of Theorem 10.1. In estimating the grid spacing h in the initial spline series with box splines M_{221}, M_{122}, or M_{1111}, the second difference operator

(10.5) $$M = -\frac{1}{2} \sum_{i,j} M_{tuvw}(i,j) \Delta_{ij}^2,$$

where Δ_{ij}^2 is the second (central) difference

$$(\Delta_{ij}^2 F)(\ell, m) = f(\ell + i, m + j) - 2f(\ell, m) + f(\ell - i, m - j),$$

with $F = \{f(\ell, m)\}$, is a main tool, and because of (10.5), it is written as a linear combination of three-second differences for the cubic splines

M_{221}, M_{122}, and two-second differences for the quadratic spline M_{1111}. The analysis for the higher order splines differs from that of the initial ones in that a different M is used to calculate higher order approximants.

We only state the following two results on convexity preservation without going into any details.

THEOREM 10.3. *Let f be a convex function in $\mathbf{R}^2$ satisfying: $f \in C^{2,1}, |D^\alpha f(\mathbf{x}) - D^\alpha f(\mathbf{y})| \leq L|\mathbf{x}-\mathbf{y}|$ for all $|\alpha| = 2$, and $D_\mathbf{u}^2 f \geq \varepsilon > 0$ for all unit vectors $\mathbf{u} \in \mathbf{R}^2$. Consider the quasi-interpolants*

$$(10.6) \qquad (P_k^h f)(\mathbf{x}) = \sum_{\mathbf{j} \in \mathbf{Z}^2} (I + M + \cdots + M^k) F_h(\mathbf{j}) M_{tuv}(\frac{1}{h}\mathbf{x} - \mathbf{j})$$

where $M(\mathbf{j}) = \delta_{0\mathbf{j}} - M_{tuv}(\mathbf{j})$, $F_h = \{f(h\mathbf{j})\}$ and $t, u \geq 2, v \geq 1$. Then $P_k^h f$ is also convex, provided that the following is satisfied:

(i) $h \leq \frac{1}{1.10L}\varepsilon$ *for $k = 0$,*

(ii) $h \leq \frac{1}{(1.10+2r)L}\varepsilon$ *for $k = 1$,*

(iii) $h \leq \frac{1}{(1.10+6r)L}\varepsilon$ *for $k = 2$, or*

(iv) $h \leq \frac{1}{(1.10+6r+A)L}\varepsilon$ *for $3 \leq k \leq \infty$*

where $r = \sum |\mathbf{i}| M_{tuv}(\mathbf{i})$ and

$$A = \frac{2a^2(2-a)}{(1-a)^2} r \left(\sum_{M_{tuv}(\mathbf{i}) \neq 0} 1 \right)^{1/2}$$

with $a = \max \widetilde{M}_{tuv}(\omega)$.

Here, $P_\infty^h f$ interpolates f at $h\mathbf{j}$, $\mathbf{j} \in \mathbf{Z}^2$. Of course, for sufficiently smooth f and appropriately large k, the approximation order of $P_k^h f$ to f is optimal.

THEOREM 10.4. *Let f be a convex function in $\mathbf{R}^2$ satisfying: $f \in C^{2,1}, |D^\alpha f(\mathbf{x}) - D^\alpha f(\mathbf{y})| \leq L|\mathbf{x}-\mathbf{y}|$ for all $|\alpha| = 2$, and $D_\mathbf{u}^2 f \geq \varepsilon > 0$ for all unit vectors $\mathbf{u}$ in $\mathbf{R}^2$. Then the quasi-interpolant in (10.6) with M_{tuv} replaced by M_{tuvw} where $t, u \geq 2$ and $v, w \geq 1$ is also convex, provided that the following is satisfied:*

(i) $h \leq \frac{1}{1.10L}\varepsilon$ *for $k = 0$,*

(ii) $h \leq \frac{1}{(1.10+2r)L}\varepsilon$ *for $k = 1$,*

(iii) $h \leq \frac{1}{(1.10+6r)L}\varepsilon$ *for $k = 2$, or*

(iv) $h \leq \frac{1}{(1.10+6r+(k+1)(k-2)rB)L}\varepsilon$ for $3 \leq k < \infty$

where $r = \sum |\mathbf{i}| M_{tuvw}(\mathbf{i})$ and

$$B = \left(\sum_{M_{tuvw}(\mathbf{i}) \neq 0} 1 \right)^{1/2}.$$

Of course, the approximation order is again optimal for appropriately smooth f and large k. As we recall from §9.4, however, the sequence $\{P_k^h f\}$ of quasi-interpolants in Theorem 10.4 does not converge since we have a four-directional mesh.

10.3. Application to CAGD. In this section, we will discuss the following important problem in CAGD: A set of discrete data $\{f_i\}$ taken at a finite number of points $\{(x_i, y_i)\}$ in $\mathbf{R}^2$, where $i = 1, \cdots, V$, is given, and one is asked to interpolate the data by a C^1 piecewise quadratic polynomial, such that the "spline" interpolant must preserve the shape characteristics of the given data.

Of course, as we have seen in §10.1, the shape characteristics of the data depend on the triangulation $\triangle$, say, of the set of nodes, which we assume has been done, and as usual, the triangulation must satisfy conditions (i) - (iii) in §4.3. Hence, the piecewise linear polynomial interpolant on this triangulation exists and determines the shape characteristics of the data. The reason for using quadratic instead of higher degree polynomials is to avoid as much as possible any oscillation.

We approach this problem by using generalized vertex splines introduced in §6.2. That is, each triangle is divided into twelve subtriangles using the medians to obtain the refinement $\widehat{\triangle} \supset \triangle$, and the space $S_2^1(\widehat{\triangle})$ of all bivariate C^1 quadratic splines on $\widehat{\triangle}$ has dimension $3V + E$, where E is the number of edges of the original triangulation $\triangle$. In fact, $S_2^1(\widehat{\triangle})$ has a basis given by S_i^*, T_i^*, U_i^* and V_j, where $i = 1, \cdots, V$ and $j = 1, \cdots, E$. Here, S_i^*, T_i^*, U_i^* are generalized 0-vertex splines defined in §6.2 and V_j is a generalized 1-vertex spline which has zero $(\delta, \delta_x, \delta_y)$ values at all vertices and zero δ_n values at all edges except the jth edge, where it has the value 1 (see §6.2 and §6.3 for more details). We have seen from Theorem 6.2 that the 1-vertex splines V_j are not needed for the spline series (6.3) to preserve all quadratic polynomials. However, in CAGD the order of approximation is not as important as the shape characteristics, and the V_j's can be used to "patch up" the shape across certain edges. The Bézier nets of S_i^*, T_i^*, U_i^* and V_j are given in §6.3. Now, the spline function

(10.7)
$$s_f(x,y) = \sum_{i=1}^{V}[f_i S_i^*(x,y) + m_i T_i^*(x,y)$$
$$+ n_i U_i^*(x,y)] + \sum_{j=1}^{E} p_j V_j(x,y)$$

clearly satisfies the required interpolation condition

$$s_f(x_i, y_i) = f_i, \quad i = 1, \cdots, V,$$

and the parameters m_i, n_i, p_j can be used to adjust the shape characteristics of s_f. In Chui and He [60] a set C of sufficient conditions on $A = \{m_i, n_i, p_j\}$ described by certain linear inequalities that guarantee the convexity of s_f is given. If the data $\{f_i\}$ on the triangulation Δ is convex, then the set of spline functions satisfying these sufficient conditions is nonempty. In the same paper, a set M of necessary and sufficient conditions on the parameters $A = \{m_i, n_i, p_j\}$ for s_f to be monotone nondecreasing in the direction of $\xi \in \mathbf{R}^2 \backslash \{\mathbf{0}\}$ is also given.

It would be ideal if some optimal criteria for choosing the parameters $A = (m_i, n_i, p_j)$ from C or M can be determined. For the convexity problem, let $H(A)$ denote the Hessian matrix of s_f. Then $H(A)$ is a piecewise constant matrix-valued function of A. If we want s_f to be convex but the "average convexity measurement" to be minimized, then it seems reasonable to consider the problem of minimizing the quantity

$$\frac{1}{2\pi} \int_0^{2\pi} [\cos t \ \sin t] H(A) \begin{bmatrix} \cos t \\ \sin t \end{bmatrix} dt$$

subject to $A \in C$. That is, we are faced with the following linear programming problem with linear constraints:

(10.8) $$\min\{\nabla^2 s_f : A \in C\}.$$

Of course, if we use a set of necessary and sufficient conditions on convexity, the constraints become nonlinear. For the monotonicity problem, if we wish to minimize the "average increment," it seems reasonabe to consider the problem of minimizing the quantity

(10.9) $$\iint_\Omega |\nabla s_f|^2 dx\, dy = \iint_\Omega \left[\left(\frac{\partial s_f}{\partial x}\right)^2 + \left(\frac{\partial s_f}{\partial y}\right)^2\right] dx dy$$

subject to $A \in M$. This is a quadratic programming problem with linear constraints. In Chui, Chui, and He [51], both of these two problems as well as the problems:
$$\min \{\nabla^2 s_f \colon \ A \in M\}$$
and
$$\min \left\{ \iint_\Omega |\nabla s_f|^2 dx\, dy \colon s_f \geq 0 \right\}$$
for the three-directional mesh $\triangle = \triangle_{MN}^{(1)}$ are studied in some detail.

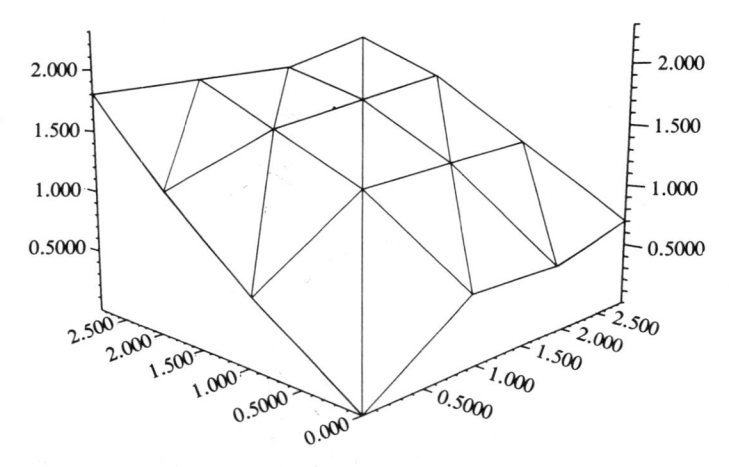

FIG. 10.2

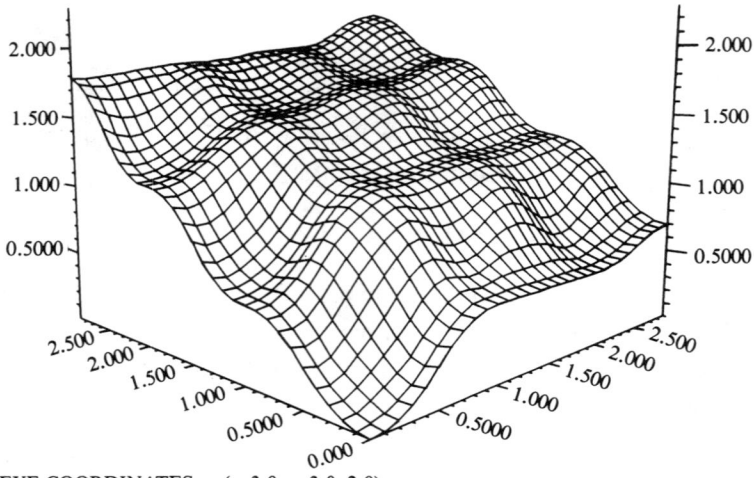

FIG. 10.3

SHAPE-PRESERVING 167

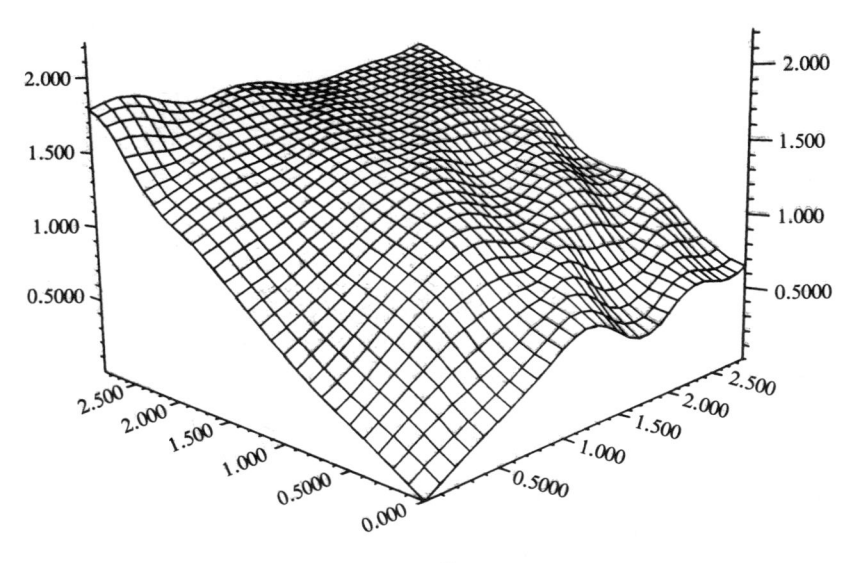

FIG. 10.4

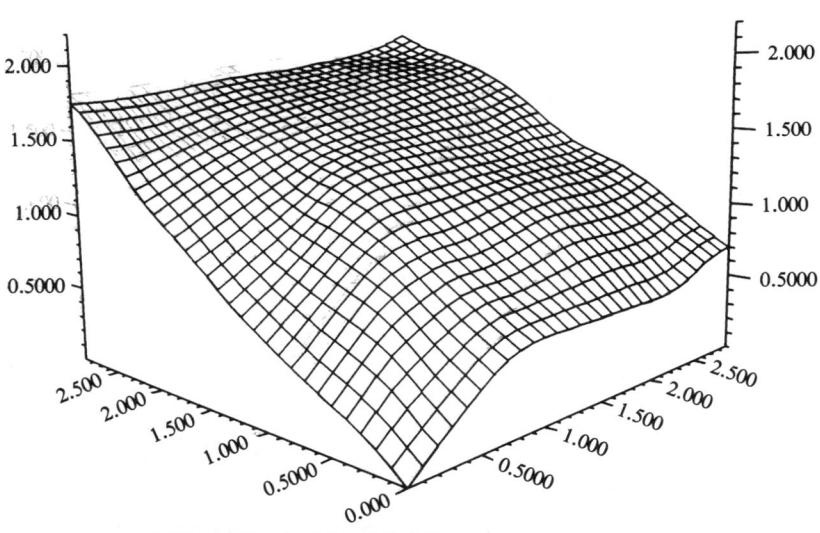

FIG. 10.5

In Figure 10.2 we give an example of a monotone discrete data set. The graph in Figure 10.3 is drawn by using zero first partial derivative values for m_i and n_i. In Figure 10.4, we give a graph by choosing m_i and n_i to be divided differences of the data values. Note that the surface in Figure 10.4 is not monotone. Our optimal monotone interpolant satisfying (10.9) with constraint $A \in M$ is shown in Figure 10.5.

10.4. Reconstruction of gradient fields. The problem to be discussed in this section can be stated as follows: Let $\mathbf{x}^i \in \Omega \subset \mathbf{R}^s$, $i = 1, \cdots, V$, and suppose that the data set $\{\mathbf{F}_i : i = 1, \cdots, V\}$ is vector-valued and represents a gradient field of some *unknown* potential function f at the nodes $\mathbf{x}^i$; that is,

$$(10.10) \qquad (\nabla f)(\mathbf{x}^i) = \mathbf{F}_i.$$

The problem is to reconstruct $(\nabla f)(\mathbf{x})$, $\mathbf{x} \in \Omega$. In the bivariate setting, this problem is posed in Chui [50] in the study of wind currents over terrains using scattered vector-valued data taken at certain sensors. The approach is to write

$$(10.11) \qquad f(x,y) = \sum_{i=1}^{V} [f_i S_i^*(x,y) + m_i T_i^*(x,y) + n_i U_i^*(x,y)]$$

where $\{\mathbf{F}_i\}$, $\mathbf{F}_i = (m_i, n_i)$, is the given data set, and the coefficients $\{f_i\}$ must be determined. Note that the interpolation condition (10.10) is automatically satisfied. To determine $\{f_i\}$, the extremal problem

$$(10.12) \qquad \min_{\{f_i\}} \int\int_\Omega (\nabla^2 f(x,y))^2 \, dxdy$$

is solved in Chui [50], where C^1 quintic vertex splines are used instead, and the computational procedure is not very complicated. The result has obviously improved over the classical approach where each component of $\{F_i\}$ is interpolated separately. Using the generalized vertex splines as in (10.11) has to simplify the computational procedure.

10.5. Applications to signal processing. Multivariate splines also find important applications to various areas in signal processing. One such area is the *foveal problems* in image processing. These problems are concerned with transmitting in high resolution the area of interest and successively lower the resolution in an area away from the defined special interest segment called the *foveal spot*. The purpose of segmentation of an image is to help define this foveal spot more accurately. Once this area has been designated and transmitted, the observer must have this special interest area magnified to normal roster size in order to allow objects to be recognized and full resolution capacity utilized. To use full resolution, it is necessary to determine, as accurately as possible, the intensity of the missing picture elements; and on the other hand, it is necessary to smooth the data so that the image presents a pleasing effect to the viewer. To apply multivariate splines, we have to study approximation of (segments of) a spline surface by one with fewer grid lines. This may be classified as data reduction. At the same time, the missing picture elements in the area of interest must be interpolated using a spline surface segment with more refined grid, so that the whole surface remains to be smooth. To transmit the picture, the Bézier nets and/or coefficients of box spline series may be used to represent the entire picture.

Another area in signal processing in which multivariate splines should have some applications is *optical filtering*. As it is well known, an optical lens transforms the input signal from the spatial domain to the frequency domain by means of Fourier transformation. This suggests an application of box splines in the filtering process since the Fourier transform of a box spline has very simple formulation.

APPENDIX

A Computational Scheme for Interpolation

Harvey Diamond
West Virginia University

This appendix describes an iterative computational scheme for interpolation on bounded domains in the nonsingular case. The schemes discussed in Chapter 9 were based on cardinal interpolation in the nonsingular case (cf. Theorem 9.11) or precomputation of the fundamental sequence Λ^* in the singular case, and neither of them, strictly speaking, is computationally feasible as they are based on solutions of cardinal interpolation problems.

Given the compact domain W, and a function $f \in C^{n_0,1}(\Omega)$, Ω open and $W \subset \Omega$, we need a set C of coefficients satisfying

$$\Phi C(\mathbf{j}) = f(h\mathbf{j}), \quad h\mathbf{j} \in W$$

and giving the optimal order $O(h^{n_0+1})$ of approximation on W. Replacing Φ by $I - M$, we may write the interpolation equations as

(A.1) $$C(\mathbf{j}) = f(h\mathbf{j}) + MC(\mathbf{j}), \quad h\mathbf{j} \in W.$$

The interpolation scheme is as follows:

Step (1). Calculate the coefficients C_q of a quasi-interpolant on W by using the iteration

(A.2) $$C^{(n)}(\mathbf{j}) = f(h\mathbf{j}) + MC^{(n-1)}(\mathbf{j}), \quad C^{(-1)} \equiv 0$$

until $n \geq (n_0 - 1)/2$.

This step requires knowledge of the values of f on a slightly larger set than $\{h\mathbf{j}: h\mathbf{j} \in W\}$, sufficient to correctly caluculate the coefficients $C_q(\mathbf{j})$ at all $\mathbf{j}$ in the set

$$\mathbf{Z}_w = \{\, \mathbf{j}: \; W \cap \operatorname{supp}(\phi(\cdot/h - \mathbf{j})) \neq \emptyset \,\};$$

that is, the set of coefficients of all scaled splines whose supports intersect W. Alternately, if the function value required for the calculation in (A.2) are not available off of W, a local extrapolation formula exact for polynomials of total degree n_0 can be used to approximate the values.

Step (2). Perform the iteration, for $\mathbf{j} \in \mathbf{Z}_w$

(A.3)
$$C^{(0)}(\mathbf{j}) = C_\alpha(\mathbf{j})$$
$$C^{(n)}(\mathbf{j}) = \begin{cases} f(h\mathbf{j}) + MC^{(n-1)}(h\mathbf{j}) & \text{if } h\mathbf{j} \in W \\ C_q(\mathbf{j}) & \text{if } h\mathbf{j} \notin W \end{cases}$$

until the desired convergence is obtained.

Note that
$$\Phi C^{(n-1)}(\mathbf{j}) = (I - M)C^{(n-1)}(\mathbf{j})$$
$$= f(h\mathbf{j}) + C^{(n-1)}(\mathbf{j}) - C^{(n)}(\mathbf{j}), \quad h\mathbf{j} \in W$$

so that the interpolation error at the nodes of W is given by

$$f(h\mathbf{j}) - \Phi C^{(n-1)}(\mathbf{j}) = C^{(n)}(\mathbf{j}) - C^{(n-1)}(\mathbf{j}), \quad h\mathbf{j} \in W$$

and the difference of two successive iterates therefore gives the interpolation error.

To explain why this scheme works, in Step (2), set $C^{(n)} = C_q + \xi^{(n)}$, giving as the recursion for $\mathbf{j} \in \mathbf{Z}_w$

$$\xi^{(0)}(\mathbf{j}) \equiv 0$$
$$\xi^{(n)}(\mathbf{j}) = \begin{cases} f(h\mathbf{j}) - \Phi C_q(\mathbf{j}) + M\xi^{(n-1)}(h\mathbf{j}) & \text{if } h\mathbf{j} \in W \\ 0 & \text{if } h\mathbf{j} \notin W. \end{cases}$$

This recursion converges to a solution of $\Phi\xi(\mathbf{j}) = f(h\mathbf{j}) - \Phi C_q(\mathbf{j})$, $h\mathbf{j} \in W$; moreover, this solution has the same order of magnitude as $f(h\mathbf{j}) - \Phi C_q(\mathbf{j})$, namely $O(h^{n_0+1})$. These results are shown in [54] and are based on the fact that the linear operator M_w defined on the sequences $\{\xi : \xi(\mathbf{j}) = 0, h\mathbf{j} \notin W\}$ by

$$M_w(\xi)(\mathbf{j}) = \begin{cases} M\xi(\mathbf{j}) & \text{if } h\mathbf{j} \in W \\ 0 & \text{otherwise} \end{cases}$$

satisfies $\| M_w(\xi) \|_2 \le r \| \xi \|_2$ where $r = 1 - \min \tilde{\phi}(\omega)$. The coefficients of the interpolant generated by the recursion (A.3) thus differ from C_q by $O(h^{n_0+1})$, which shows that the interpolant has the optimal order of approximation.

The algorithm was applied to the interpolation of the function $f(x,y) = y \sin x$ on the domain $(-6.2, 6.2) \times (-1,1)$ using the box spline M_{222} which is C^2 with order of approximation $O(h^4)$. The values of M_{222} at the integers are $1/2$ at $(0,0)$ and $1/12$ at the points $(\pm 1, 0), (0, \pm 1)$, and $\pm(1,1)$. The coefficients of the quasi-interpolant are given by

APPENDIX 173

$$c_{ij} = (I + M)F_h(i,j).$$

To implement the algorithm, data at the nodes of W needs to be augmented by roughly two layers.

The graphs on the following pages are as follows:

(1) Graph of the function $f(x,y) = y \sin x$ on the domain $(-6.2, 6.2) \times (-1, 1)$ with points plotted with spacing $12.4/80$ in the x-direction and $1/40$ in the y-direction (see Figure A.1).

(2) The data points are spaced by $12.4/20$ in the x-direction and $1/10$ in the y-direction. A graph of the data is shown (see Figure A.2).

(3) The quasi-interpolant calculated from the data is plotted at the same set of points as the original function. This graph is virtually indistinguishable from the original function (see Figure A.3).

(4) The interpolation error of the quasi-interpolant is graphed at the data points (see Figure A.4).

(5) The approximation error of the quasi-interpolant over the domain is graphed (see Figure A.5).

(6) The interpolation error using the approximant obtained two iterations after the quasi-interpolant is graphed. For M_{222}, $r = 3/4$ so that the interpolation error should decrease with each iteration approximately by a factor of $3/4$; the error in this graph is roughly $1/10$ the size of that in (4) (see Figure A.6).

(7) The approximation error over the domain using the approximant obtained two iterations after the quasi-interpolant (see Figure A.7).

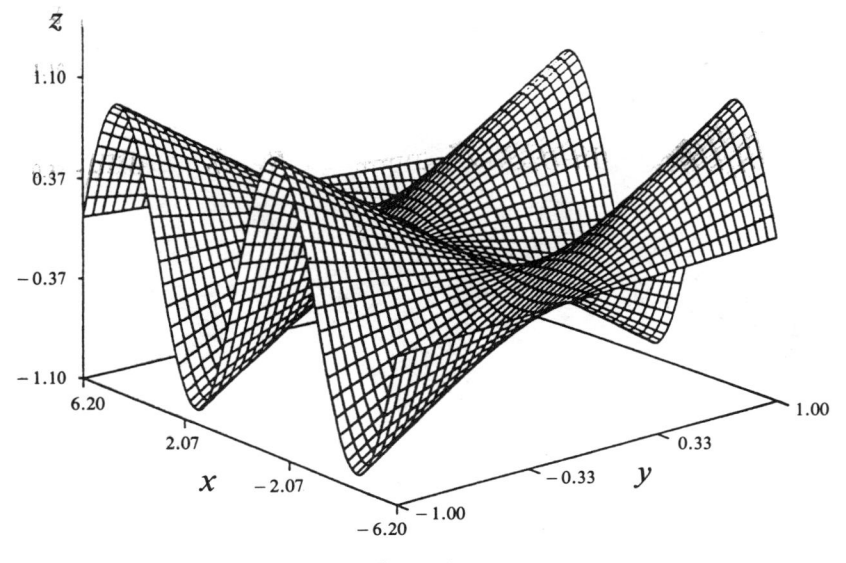

FIG. A.1

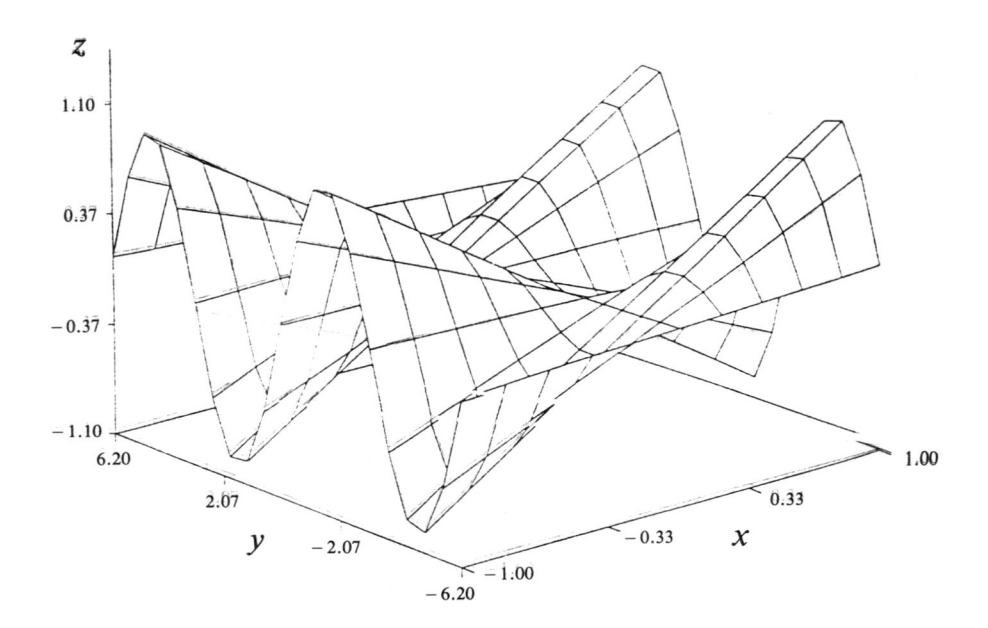

FIG. A.2

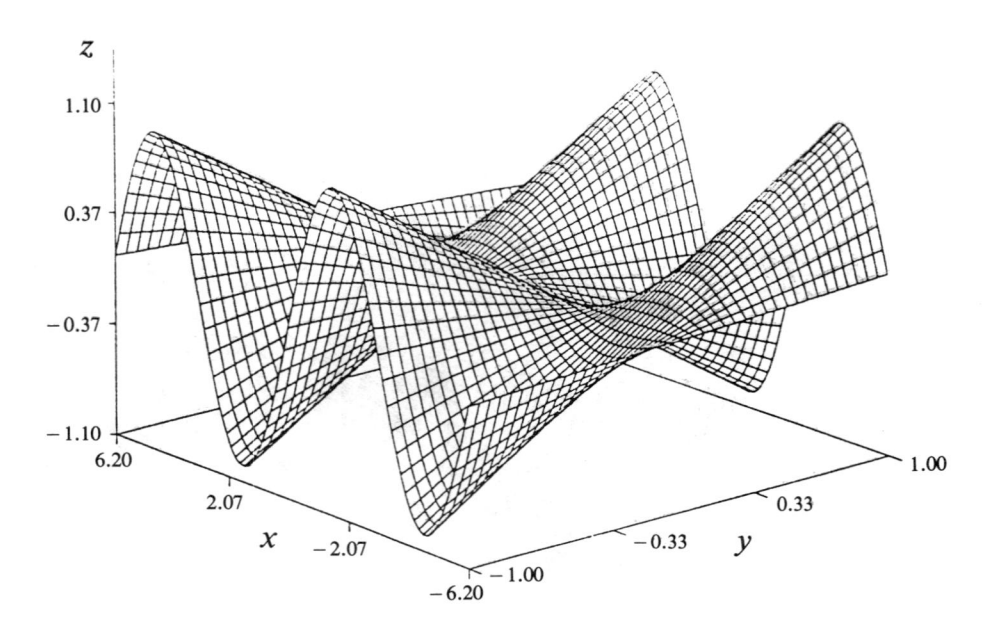

FIG. A.3

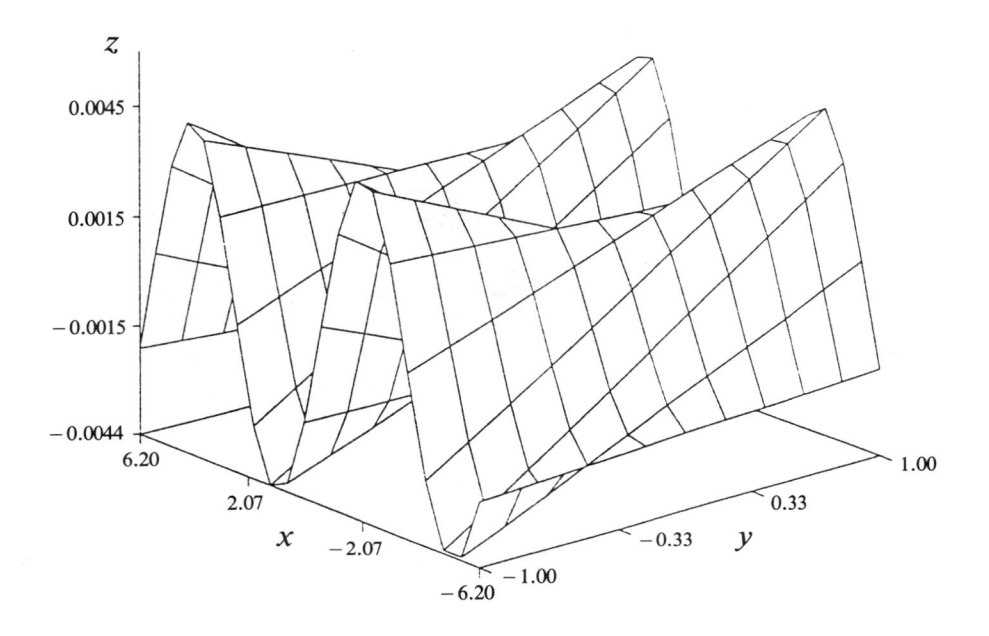

FIG. A.4

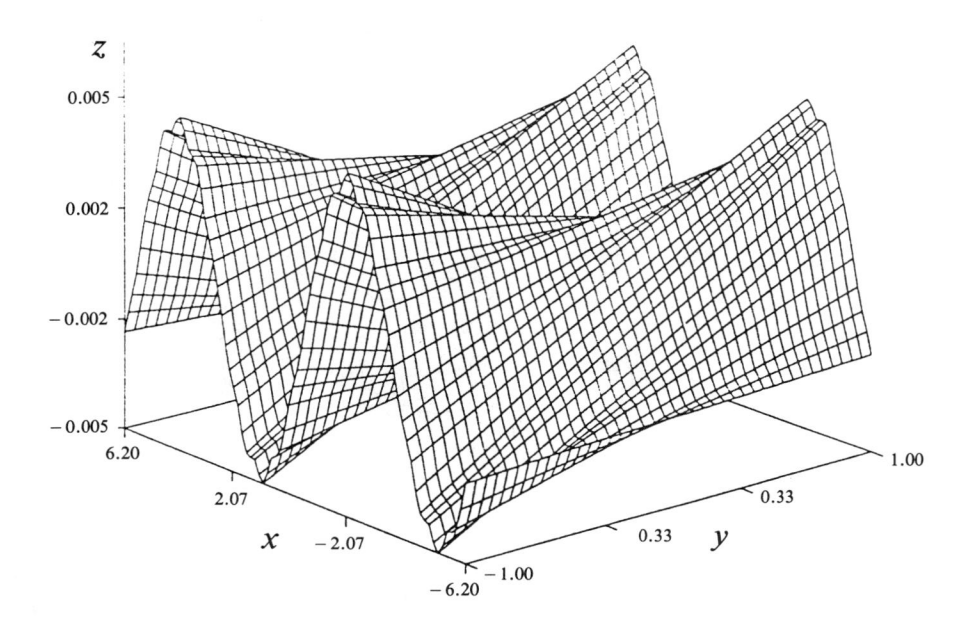

FIG. A.5

APPENDIX

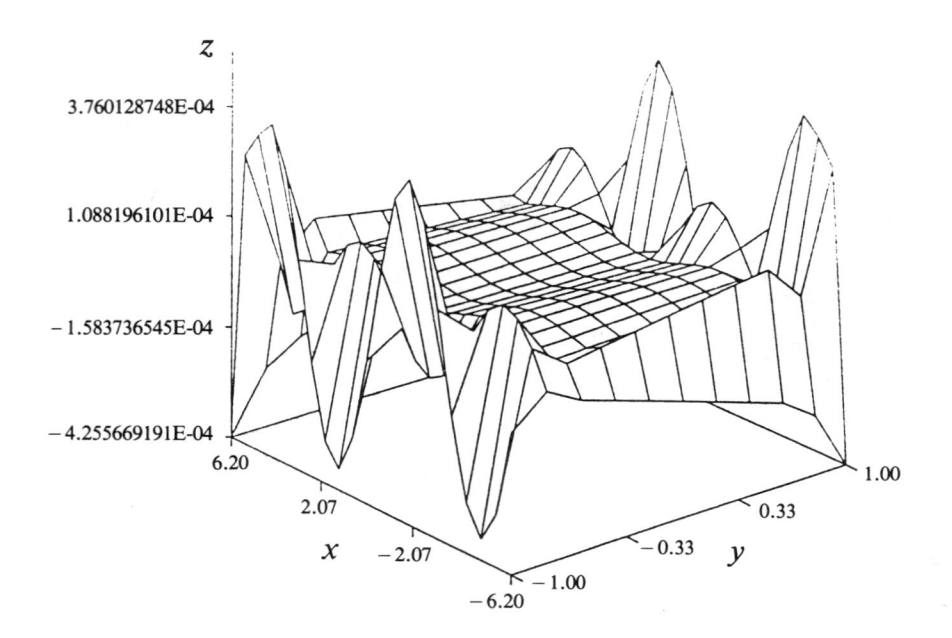

FIG. A.6

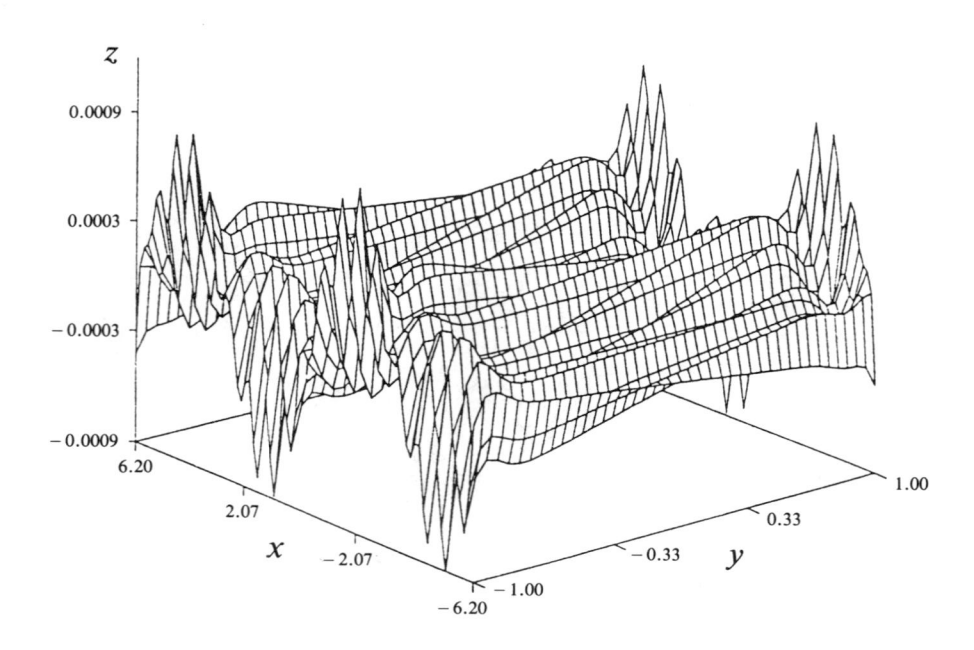

FIG. A.7

Bibliography

[1] ALFELD, P., *A discrete C^1 interpolant for tetrahedral data*, Rocky Mountain J. Math., 14 (1984), pp. 5–16.

[2] ———, *Derivative generation from multivariate scattered data by functional minimization*, Comput. Aided Geom. Design, 2 (1985), pp. 281–296.

[3] ———, *On the dimension of multivariate piecewise polynomials*, in Proceedings of the Biennial Dundee Conference on Numerical Analysis, Pitman, Harlow, Essex, 1985.

[4] ALFELD, P., B. PIPER AND L. L. SCHUMAKER, *Minimally supported bases for spaces of bivariate piecewise polynomials of smoothness r and degree $d \geq 4r + 1$*, Comput. Aided Geom. Design, 3 (1987), pp. 189–198.

[5] ———, *Spaces of bivariate splines on triangulations with holes*, Approx. Theory Appl., to appear.

[6] ALFELD, P. AND L. L. SCHUMAKER, *The dimension of bivariate spline spaces of smoothness r for degree $d \geq 4r + 1$*, Constr. Approx., 3 (1987), pp. 189–197.

[7] BAKER, B. S., E. GROSSE, AND C. S. RAFFERTY, *Non-obtuse triangulation of a polygon*, Bell Lab Rpt., Murray Hill, 1985.

[8] BAMBERGER, L., *Zweidimensionale Splines auf regulären Triangulationen*, Ph.D. Thesis, University of Munich, 1985.

[9] BARNHILL, R. E. AND G. FARIN, *C^1 quintic interpolation over triangles: two explicit representations*, J. Numer. Meth. Engr., 17 (1981), pp. 1763–1778.

[10] BEATSON, R. AND Z. ZIEGLER, *Monotonicity preserving surface interpolation*, SIAM J. Numer. Anal., 22 (1985), pp. 401–411.

[11] BÉZIER, P., *Emploi des Machines à Commande Numérique*, Masson et Cie, Paris, 1970.

[12] ———, *Numerical Control-Mathematics and Applications*, John Wiley, London, 1972.

[13] BÉZIER, P., *Mathematical and practical possibilities of* UNISURF, in Computer Aided Geometric Design, R.E. Barnhill and R.F. Riesenfeld, eds., Academic Press, New York, 1974, 127–152.

[14] ———, *A view of* CAD-CAM, Computer-aided Design, 13 (1981), pp. 207–209.

[15] BILLERA, L. J., *Homology of smooth splines: Generic triangulations and a conjecture of Strang*, unpublished manuscript.

[16] BÖHM, W., *Subdividing multivariate splines*, Computer-aided Design, 15 (1983), pp. 345-352.

[17] BÖHM, W., G. FARIN, AND J. KAHMANN, *A survey of curve and surface methods in* CAGD, Comput. Aided Geom. Design, 1 (1984), pp. 1–60.

[18] DE BOOR, C., *On uniform approximation by splines*, J. Approx. Theory, 1 (1968), pp. 219-235.

[19] ———, *Splines as linear combinations of B-splines, a survey*, in Approximation Theory II, G. G. Lorentz, C. K. Chui, and L. L. Schumaker, eds., Academic Press, New York, 1976, pp. 1–47.

[20] ———, *A Practical Guide to Splines*, Springer-Verlag, New York, 1978.

[21] ———, *Topics in multivariate approximation theory*, in Topics in Numerical Analysis, P. Turner, ed., Lecture Notes 965, Springer-Verlag, Berlin, 1982, pp. 39–78.

[22] ———, *B-form basics*, in Geometric Modeling, G. Farin, ed., Society for Industrial and Applied Mathematics, Philadelphia, 1987, pp. 131–148.

[23] ———, *The polynomials in the linear span of integer translates of a compactly supported function*, Constr. Approx., 3 (1987), pp. 199–208.

[24] ———, *Multivariate approximation*, in State of the Art in Numerical Analysis, A. Iserles and M. Powell, eds., Institute Mathematics Applications, Essex, 1987.

[25] DE BOOR, C. AND R. DEVORE, *Approximation by smooth multivariate splines*, Trans. Amer. Math. Soc., 276 (1983), pp. 775–788.

[26] DE BOOR, C., R. DEVORE, AND K. HÖLLIG, *Approximation order from smooth bivariate pp functions*, in Approximation Theory IV, C. K. Chui, L. L. Schumaker, and J. D. Ward, eds., Academic Press, New York, 1983, pp. 353–357.

[27] DE BOOR, C. AND G. J. FIX, *Spline approximation by quasi-interpolants*, J. Approx. Theory, 8 (1973), pp. 19–45.

[28] DE BOOR, C. AND K. HÖLLIG, *Recurrence relations for multivariate B-splines*, Proc. Amer. Math. Soc., 85 (1982), pp. 397–400.

[29] ———, *B-splines from parallelepipeds*, J. Analyse Math., 42 (1982), pp. 99–115.

[30] ———, *Approximation order from bivariate C^1-cubics: a counterexample*,

Proc. Amer. Math. Soc., 87 (1983), pp. 649–655.

[31] DE BOOR, C. AND K. HÖLLIG, *Bivariate box splines and smooth pp functions on a three-direction mesh*, J. Comput. Appl. Math., 9 (1983), pp. 13–28.

[32] ———, *Approximation power of smooth bivariate pp functions*, MRC Rpt. 2967, University of Wisconsin, Madison, 1987.

[33] ———, *Minimal support for bivariate splines*, Approx. Theory Appl., 3 (1987), pp. 11–23.

[34] DE BOOR, C., K. HÖLLIG, AND S. RIEMENSCHNEIDER, *Bivariate cardinal interpolation*, in Approximation Theory IV, C. K. Chui, L. L Schumaker, and J. D Ward, eds., Academic Press, New York, 1983, pp. 359–363.

[35] ———, *On bivariate cardinal interpolation*, in Constructive Theory of Functions '84, B. Sendov, P. Petrushev, R. Maleev, and S. Tashev, eds., Bulgarian Academy of Sciences, Sofia, 1984, pp. 254–259.

[36] ———, *Bivariate cardinal interpolation by splines on a three-direction mesh*, Illinois J. Math., 29 (1985), pp. 533–566.

[37] ———, *The limits of multivariate cardinal splines*, in Multivariate Approximation Theory III, W. Schempp & K. Zeller, eds., Birkhäuser, Basel, 1985, pp. 47–50.

[38] ———, *Convergence of bivariate cardinal interpolation*, Constr. Approx., 1 (1985), pp. 183–193.

[39] ———, *Convergence of cardinal series*, Proc. Amer. Math. Soc., to appear.

[40] DE BOOR, C. AND R. Q. JIA, *Controlled approximation and a characterization of the local approximation order*, Proc. Amer. Math. Soc., 95 (1985), pp. 547–553.

[41] BRAMBLE, J. AND M. ZLAMAL, *Triangular elements in the finite element method*, Math. Comp., 24 (1970), pp. 809–820.

[42] CARLSON, R. E. AND F. N. FRITSCH, *Monotone piecewise bicubic interpolation*, SIAM J. Numer. Anal., 22 (1985), pp. 386–400.

[43] CHANG, G. AND P. J. DAVIS, *The convexity of Bernstein polynomials over triangles*, J. Approx. Theory, 40 (1984), pp. 11–28.

[44] CHANG, G. AND Y. FENG, *An improved condition for the convexity of Bernstein-Bézier surfaces over triangles*, Comput. Aided Geom. Design, 1 (1984), pp. 279–283.

[45] ———, *A new proof for the convexity of the Bernstein-Bézier surfaces over triangles*, Chinese. Ann. Math. Ser. B, 6 (1985), pp. 173–176.

[46] CHENEY, E. W., *Multivariate Approximation Theory: Selected topics*, CBMS Vol. 51, Society for Industrial and Applied Mathematics, Philadelphia, 1986.

[47] CHUI, C. K., *Approximations and expansions*, Ency. of Physical Sciences and Technology, Academic Press, New York, 1987, pp. 661–687.

[48] CHUI, C. K., *Bivariate quadratic splines on crisscross triangulations*, Trans. First Army Conf. Appl. Math. Comp., 1 (1984), pp. 877–882.

[49] ———, *B-splines on nonuniform triangulations*, Trans. Second Army Conf. Appl. Math. Comp., 2 (1985), pp. 939–942.

[50] ———, *Multi-dimensional spline techniques for fitting of surface to wind fields over complex terrain*, Final Report, Battelle, Aug. 1986.

[51] CHUI, C. K., H. C. CHUI, AND T. X. HE, *Shape-preserving interpolation by bivariate C^1 quadratic splines*, CAT 148, Texas A&M University, 1987.

[52] CHUI, C. K. AND H. DIAMOND, *A natural formulation of quasi-interpolation by multivariate splines*, Proc. Amer. Math. Soc., 99 (1987), pp. 643–646.

[53] CHUI, C. K., H. DIAMOND, AND L. A. RAPHAEL, *Interpolation by bivariate quadratic splines on nonuniform rectangles*, Trans. Fourth Army Conf. Appl. Math. Comp., 87- 1 (1987), pp. 1261–1266.

[54] ———, *Interpolation by multivariate splines*, Math. Comp., to appear.

[55] ———, *Shape-preserving quasi-interpolation and interpolation by box spline surfaces*, CAT 146, Texas A&M University, 1987.

[56] CHUI, C. K. AND T. X. HE, *On minimal and quasi-minimal supported bivariate splines*, J. Approx. Theory, 52 (1988), pp. 217–238.

[57] ———, *Computation of minimal and quasi-minimal supported bivariate splines*, J. Comp. Math., to appear.

[58] ———, *On location of sample points in C^1 quadratic bivariate spline interpolation*, in Numerical Methods of Approximation Theory, Vol.8, L. Collatz, G. Meinardus, & G. Nürnberger, eds., Birkhäuser, Basel, 1987, pp. 30–43.

[59] ———, *On the dimension of bivariate super spline spaces*, CAT 144, Texas A&M University, 1987.

[60] ———, *On bivariate C^1 quadratic finite elements and vertex splines*, CAT 147, Texas A&M University, 1987.

[61] CHUI, C. K., T. X. HE, AND R. H. WANG, *Interpolation by bivariate linear splines*, in Alfred Haar Memorial Conference, J. Szabados and K. Tandori, eds., North-Holland, Amsterdam, 1986, pp. 247–255.

[62] ———, *The C^2 quartic spline space on a four-directional mesh*, Approx. Theory Appl., 3 (1987), pp. 32–36.

[63] CHUI, C. K., K. JETTER, AND J. D. WARD, *Cardinal interpolation by multivariate splines*, Math. Comp., 48 (1987), pp. 711–724.

[64] CHUI, C. K. AND M. J. LAI, *On bivariate vertex splines*, in Multivariate Approximation Theory III, W. Schempp & K. Zeller, eds., Birkhäuser, Basel, 1985, pp. 84–115.

[65] ———, *A multivariate analog of Marsden's identity and a quasi-interpolation scheme*, Constr. Approx., 3 (1987), pp. 111–122.

[66] CHUI, C. K. AND M. J. LAI, *On multivariate vertex splines and applications*, in Topics in Multivariate Approximation, C. K. Chui, L. L. Schumaker, and F. Utreras, eds., Academic Press, New York, 1987, pp. 19–36.

[67] ———, *Computation of box splines and B-splines on triangulations of nonuniform rectangular partitions*, Approx. Theory Appl., 3 (1987), pp. 37–62.

[68] ———, *Vandermonde determinants and Lagrange interpolation in $\mathbb{R}^s$*, in Nonlinear and Convex Analysis, B. L. Lin and S. Simons, eds., Marcel Dekker, New York, 1987, pp. 23-36.

[69] CHUI, C. K. AND L. L. SCHUMAKER, *On spaces of piecewise polynomials with boundary conditions*, I. Rectangles, in Multivariate Approximation Theory II, W. Schempp & K. Zeller, eds., Birkhäuser, Basel, 1982, pp. 69–80.

[70] CHUI, C. K., L. L. SCHUMAKER, AND F. UTRERAS, eds., *Topics in Multivariate Approximation*, Academic Press, New York, 1987.

[71] CHUI, C. K., L. L. SCHUMAKER, AND R. H. WANG, *On spaces of piecewise polynomials with boundary conditions*, II. Type-1 triangulations, in Second Edmonton Conference on Approximation Theory, Z. Ditzian, A. Meir, S. Riemenschneider, and A. Sharma, eds., American Mathematical Society, Providence, 1983, pp. 51–66.

[72] ———, *On spaces of piecewise polynomials with boundary conditions*, III. Type-2 triangulations, in Second Edmonton Conference on Approximation Theory, Z. Ditzian, A. Meir, S. Riemenschneider, and A. Sharma, eds., American Mathematical Society, Providence, 1983, pp. 67–80.

[73] CHUI, C. K.AND R. H. WANG, *Multivariate spline spaces*, J. Math. Anal. Appl., 94 (1983), pp. 197–221.

[74] ———, *On smooth multivariate spline functions*, Math. Comp., 41 (1983), pp. 131–142.

[75] ———, *Spaces of bivariate cubic and quartic splines on type-1 triangulations*, J. Math. Anal. Appl., 101 (1984), pp. 540–554.

[76] ———, *On a bivariate B-spline basis*, Scientia Sinica, 27 (1984), pp. 1129–1142.

[77] ———, *Concerning C^1 B-splines on triangulations of non-uniform rectangular partitions*, Approx. Theory Appl., 1 (1984), pp. 11–18.

[78] CHUNG, K. C.AND T. H. YAO, *On lattices admitting unique Lagrange interpolations*, SIAM J. Numer. Anal., 14 (1977), pp. 735–741.

[79] COHEN, E., T. LYCHE, AND R. RIESENFELD, *Discrete box splines and refinement algorithms*, Comput. Aided Geom. Design, 1 (1984), pp. 131–141.

[80] ———, *Cones and recurrence relations for simplex splines*, Constr. Approx., 3 (1987), pp. 131–142.

[81] COHEN, E. AND L. L. SCHUMAKER, *Rates of convergence of control polygons*, Comput. Aided Geom. Design, 2 (1985), pp. 229–235.

[82] DAHMEN, W., *Multivariate B-splines—Recurrence relations and linear combinations of truncated powers*, in Multivariate Approximation Theory, W. Schempp & K. Zeller, eds., Birkhäuser, Basel, 1979, pp. 64–82.

[83] ———, *On multivariate B-splines*, SIAM J. Numer. Anal., 17 (1980), pp. 179–190.

[84] ———, *Bernstein-Bézier representation of polynomial surfaces*, in Extension of B-spline Curve Algorithms to Surfaces, Siggraph 86, organized by C. de Boor, Dallas, 1986.

[85] ———, *Subdivision algorithms converge quadratically*, J. Comput. Appl. Math., 16 (1986), pp. 145–158.

[86] DAHMEN, W., N. DYN, AND D. LEVIN, *On the convergence rates of subdivision algorithms for box spline surfaces*, Constr. Approx., 1 (1985), pp. 305–322.

[87] DAHMEN, W. AND C. A. MICCHELLI, *On the linear independence of multivariate B-splines I. Triangulations of simploids*, SIAM J. Numer. Anal., 19 (1982), pp. 992–1012.

[88] ———, *On the linear independence of multivariate B-splines II. Complete configurations*, Math. Comp., 41 (1983), pp. 141–164.

[89] ———, *Translates of multivariate splines*, Linear Algebra Appl., 52 (1983), pp. 217–234.

[90] ———, *Recent progress in multivariate splines*, in Approximation Theory IV, C. K. Chui, L. L. Schumaker, and J. D. Ward, eds., Academic Press, New York, 1983, pp. 27–121.

[91] ———, *Multivariate splines—A new constructive approach*, in Surfaces in Computer Aided Design, R. Barnhill and W. Boehm, eds., North-Holland, Amsterdam, 1983, pp. 191–215.

[92] ———, *Subdivision algorithms for the generation of box-spline surfaces*, Comput. Aided Geom. Design, 1 (1984), pp. 115–129.

[93] ———, *On the approximation order from certain multivariate spline spaces*, J. Austral. Math. Soc. Ser. B, 26 (1984), pp. 233–246.

[94] ———, *On the optimal approximation rates for criss-cross finite element spaces*, J. Comput. Appl. Math., 10 (1984), pp. 255–273.

[95] ———, *Some results on box splines*, Bull. Amer. Math. Soc., 11 (1984), pp. 147–150.

[96] ———, *Line average algorithm: a method for the computer generation of smooth surfaces*, Comput. Aided Geom. Design, 2 (1985), pp. 77–85.

[97] ———, *On the solution of certain systems of partial difference equations and linear dependence of translates of box splines*, Trans. Amer. Math. Soc.,

292 (1985), pp. 305–320.

[98] DAHMEN, W. AND C. A. MICCHELLI, *On the local linear independence of translates of a box spline*, Studia Math., 82 (1985), pp. 243–262.

[99] ———, *Algebraic properties of discrete box splines*, Constr. Approx., 3 (1987), pp. 209–221.

[100] ———, *Convexity of multivariate Bernstein polynomials and box spline surfaces*, Studia Sci. Math. Hungar., to appear.

[101] DUCHON, J., *Fonctions-spline homogènes de plusieurs variables*, Thèse, Grenoble, 1980.

[102] DYN, N., *Interpolation of scattered data by radial functions*, in Topics in Multivariate Approximation, C. K. Chui, L. L. Schumaker, and F. I. Utreras, eds., Academic Press, New York, 1987, pp. 47–62.

[103] DYN, N., T. GOODMAN, AND C. MICCHELLI, *Positive powers of certain conditionally negative definite matrices*, Nederl. Akad. Wetensch. Proc. Ser. A, 89 (1986), pp. 163–178.

[104] DYN, N., D. LEVIN, AND S. RIPPA, *Numerical procedures for global surface fitting of scattered data by radial functions*, SIAM J. Sci. Stat. Comp., 7 (1986), pp. 639–659.

[105] FARIN, G., *Subsplines über Dreiecken*, Ph.D. thesis, Braunschweig, 1979.

[106] ———, *Designing C^1 surfaces consisting of triangular cubic patches*, Computer-aided Design, 14 (1982), pp. 253–256.

[107] ———, *A construction for the visual C^1 continuity of polynomial surface patches*, Computer Graphics Image Proc., 20 (1982), pp. 272–282.

[108] ———, ed., *Geometric Modeling*, Society for Industrial and Applied Mathematics, Philadelphia, 1987.

[109] FIX, G. AND G. STRANG, *Fourier analysis of the finite element method in Ritz–Galerkin theory*, Studies in Appl. Math., 48 (1969), pp. 265–273.

[110] FRANKE, R., *Recent advances in the approximation of surfaces from scattered data*, in Topics in Multivariate Approximation, C. K. Chui, L. L. Schumaker, and F. I. Utreras, eds., Academic Press, New York, 1987, pp. 79–98.

[111] FREDERICKSON, P. O., *Triangular spline interpolation*, Rpt. 6–70, Lakehead University, Thunder Bay, Ontario, 1970.

[112] ———, *Quasi-interpolation, extrapolation, and approximation on the plane*, Conf. Numerical Math., Winnipeg, 1971, pp. 159–167.

[113] ———, *Generalized triangular splines*, Rpt. 7–71, Lakehead University, Thunder Bay, Ontario, 1971.

[114] FRITSCH, F. N. AND R. E. CARLSON, *Monotonicity preserving bicubic interpolation: a progress report*, Comput. Aided Geom. Design, 2 (1985), pp. 117–121.

[115] GASCA, M. AND A. LOPEZ-CARMONA, *A general recurrence interpolation formula and its applications to multivariate interpolation*, J. Approx. Theory, to appear.

[116] GASCA, M. AND J. I. MAEZTU, *On Lagrange and Hermite interpolation in $\mathbb{R}^k$*, Numer. Math., 39 (1982), pp. 1–14.

[117] GASCA, M. AND V. RAMIREZ, *Interpolation systems in $\mathbb{R}^k$*, J. Approx. Theory, 42 (1984), pp. 36–51.

[118] GMELIG MEYLING, R. H. J., *Approximation by cubic C^1-splines on arbitrary triangulations*, Numer. Math., 51 (1987), pp. 65–85.

[119] GMELIG MEYLING, R. H. J. AND P. R. PFLUGER, *On the dimension of the spline space $S_2^1(\Delta)$ in special cases*, in Multivariate Approximation Theory III, W. Schempp & K. Zeller, eds., Birkhäuser, Basel, 1985, pp. 180–190.

[120] GOODMAN, T. N. T., *Interpolation in minimum semi-norm and multivariate B-splines*, J. Approx. Theory, 33 (1981), pp. 248–263.

[121] ———, *Shape preserving approximation by polyhedral splines*, in Multivariate Approximation Theory III, W. Schempp & K. Zeller, eds., Birkhäuser, Basel, 1985, pp. 198–205.

[122] ———, *Some properties of bivariate Bernstein-Schoenberg operators*, Constr. Approx., 3 (1987), pp. 123–130.

[123] GOODMAN, T. N. T. AND S. L. LEE, *Cardinal interpolation by D^M-splines*, Proc. Royal Soc. Edinburg, Sect A., 94 (1983), pp. 149–161.

[124] GRANDINE, T., *The computational cost of simplex spline functions*, MRC Rpt. 2926, University of Wisconsin, Madison, 1986.

[125] HAKOPIAN, H., *On multivariate spline functions, B-spline basis and polynomial interpolations*, SIAM J. Numer. Anal., 18 (1982), pp. 510–517.

[126] ———, *Interpolation by polynomials and natural splines on normal lattices*, in Multivariate Approximation Theory III, W. Schempp & K. Zeller, eds., Birkhäuser, Basel, 1985, pp. 218–220.

[127] HARDY, R. L., *Multiquadric equations of topography and other irregular surfaces*, J. Geophys. Res., 76 (1971), pp. 1905–1915.

[128] HAYES, J. G., *Fitting data in more than one variable*, in Numerical Approximation to Functions and Data, J. G. Hayes, ed., Athlone Press, London, 1970, pp. 84–97.

[129] HEINDL, G., *Über verallgemeinerte Stammfunktionen und LC-Funktionen in $\mathbb{R}^n$*, Ph.D. thesis, University of Munich, 1968.

[130] ———, *Spline-Funktionenals Interpolationsfunktionen mit Betragsminimalen n-ten Ableitungen und die Approximation von Peanofunktionalen*, Z. Angew. Math. Mech., 53 (1973), T-161–162.

[131] ———, *Interpolation and approximation by piecewise quadratic C^1-functions*

of two variables, in Multivariate Approximation Theory, W. Schempp & K. Zeller, eds., Birkhäuser, Basel, 1979, pp. 146–161.

[132] HÖLLIG, K., *Multivariate splines*, SIAM J. Numer. Anal., 19 (1982), pp. 1013–1031.

[133] _____, *Box Splines*, in Approximation Theory V, C. K. Chui, L. L. Schumaker, and J. D. Ward, eds., Academic Press, New York, 1986, pp. 71–95.

[134] HÖLLIG, K., M. MARSDEN, AND S. RIEMENSCHNEIDER, *Bivariate cardinal interpolation on the 3-direction mesh: ℓ^p-data*, unpublished manuscript.

[135] HONG, D., *On the dimensions of bivariate spline spaces*, Master dissertation, Zhejiang University, Hangzhou, 1987.

[136] JACKSON, I. R. H., *Convergence properties of radial basis functions*, DAMTP Report, Cambridge University, Cambridge, 1986.

[137] JETTER, K., *Some contributions to bivariate interpolation and cubature*, in Approximation Theory IV, C. K. Chui, L. L. Schumaker, and J. D. Ward, eds., Academic Press, New York, 1983, pp. 533–538.

[138] _____, *A short survey on cardinal interpolation by box splines*, in Topics in Multivariate Approximation, C. K. Chui, L. L. Schumaker, and F. I. Utreras, eds., Academic Press, New York, 1987, pp. 125–139.

[139] JETTER, K. AND S. RIEMENSCHNEIDER, *Cardinal interpolation with box splines on submodules of $\mathbb{Z}^d$*, in Approximation Theory V, C. K. Chui, L. L. Schumaker, and J. D. Ward, eds., Academic Press, New York, 1986, pp. 403–406.

[140] _____, *Cardinal interpolation, submodules, and the 4-direction mesh*, Constr. Approx., 3 (1987), pp. 169–188.

[141] JIA, R. Q., *On the linear independence of translates of box splines*, J. Approx. Theory, 40 (1984), pp. 158–160.

[142] _____, *Local linear independence of the translates of a box splines*, Constr. Approx., 1 (1985), pp. 175–182.

[143] _____, *Approximation order from certain spaces of smooth bivariate splines on a three-direction mesh*, Trans. Amer. Math. Soc., 295 (1986), pp. 199-212.

[144] LEE, S. L., *Fourier transforms of B-splines and fundamental splines for cardinal Hermite interpolations*, Proc. Amer. Math. Soc., 57 (1976), pp. 291–296.

[145] LE MEHAUTÉ, A., *On Hermite elements of class C^q in $\mathbb{R}^3$*, in Approximation Theory IV, C. K. Chui, L. L. Schumaker, and J. D. Ward, eds., Academic Press, New York, 1983, pp. 581–586.

[146] _____, *Interpolation et approximation par des fonctions polynomiales par morceaux dans $\mathbb{R}^n$*, Thèse, Université de Rennes, 1984.

[147] _____, *Piecewise polynomial interpolation in $\mathbb{R}^n$: basic aspects of the finite*

element method, CAT 113, Texas A&M University, 1986.

[148] LE MEHAUTÉ, A., *Unisolvent interpolation in $I\!R^n$ and the simplicial polynomial finite element method,* in Topics in Multivariate Approximation, C. K. Chui, L. L. Schumaker, and F. I. Utreras, eds., Academic Press, New York 1987, pp. 141–151.

[149] LORENTZ, G. G. AND R. A. LORENTZ, *Multivariate interpolation,* in Rational Approximation and interpolation, P. R. Graves-Morris, E. B. Saff, and R. S. Varga, eds., Springer-Verlag, Berlin, 1985, pp. 136–144.

[150] LORENTZ, R. A., *Some regular problems of bivariate interpolation,* in Constructive Theory of Functions '84, B. Sendov, P. Petrushev, R. Maleev, and S. Tashev, eds., Bulgarian Academy of Sciences, Sofia, 1984, pp. 549–562.

[151] LYCHE, T. AND L. L. SCHUMAKER, *Local spline approximation methods,* J. Approx. Theory, 15 (1975), pp. 294–325.

[152] MADYCH, W. R. AND S. A. NELSON, *Multivariate interpolation: a variational theory,* unpublished manuscript.

[153] MARSDEN, M. J., *An identity for spline functions with applications to variation-diminishing spline approximation,* J. Approx. Theory, 3 (1970), pp. 7–49.

[154] MEINGUET, J., *From Dirac distributions to multivariate representation formulas,* in Approximation Theory and Applications, Z. Ziegler, ed., Academic Press, New York, 1981, pp. 225–248.

[155] _____, *Surface spline interpolation: basic theory and computational aspects,* in Approximation Theory and Spline Functions, S. P. Singh, J. H. W. Burry, and B. Watson, eds., Reidel, Dordrecht, 1984, pp. 127–142.

[156] MICCHELLI, C. A., *A constructive approach to Kergin interpolation in $I\!R^k$: multivariate B-splines and Lagrange interpolation,* Rocky Mountain J. Math., 10 (1982), pp. 485–497.

[157] _____, *Interpolation of scattered data: distance matrices and conditionally positive definite functions,* in Approximation Theory and Spline Functions, S. P. Singh, J. H. W. Burry, and B. Watson, eds., Reidel, Dordrecht, 1984, pp. 143–145.

[158] _____, *Interpolation of scattered data: distance matrices and conditionally positive definite functions,* Const. Approx., 2 (1986), pp. 11–22.

[159] _____, *Subdivision algorithms for curves and surfaces,* in Extension of B-spline Curve Algorithms to Surfaces, Siggraph 86, organized by C. de Boor, Dallas, 1986.

[160] MORGAN, J. AND R. SCOTT, *A nodal basis for C^1 piecewise polynomials in two variables,* Math. Comp., 29 (1975), pp. 736–740.

[161] MORSCHE, H. G. TER, *Attenuation factors and multivariate periodic spline*

interpolation, in Topics in Multivariate Approximation, C. K. Chui, L. L. Schumaker, and F. I. Utreras, eds., Academic Press, New York, 1987, pp. 165–174.

[162] POWELL, M. J. D., *Piecewise quadratic surface fitting for contour plotting*, in Software for Numerical Mathematics, D. J. Evans, ed., Academic Press, London, 1974, pp. 253–271.

[163] _____, *Radial basis functions for multivariable interpolation: a review*, in Algorithms for the Approximation of Functions and Data, M. G. Cox and J. C. Mason, eds., Oxford University Press, 1987.

[164] POWELL, M. J. D. AND M. A. SABIN, *Piecewise quadratic approximations on triangles*, ACM Trans. Math. Software, 3 (1977), pp. 316–325.

[165] PRAUTZSCH, H., *Unterteilungsalgorithmen für multivariate Splines*, Ph.D. thesis, University of Braunschweig, 1984.

[166] _____, *Generalized subdivision and convergence*, Comput. Aided Geom. Design, 2 (1985), pp. 69–75.

[167] RENKA, R. J., *Triangulation and bivariate interpolation for irregularly distributed data points*, Ph.D. thesis, University of Texas, Austin, 1981.

[168] _____, *Algorithm* 623, ACM TOMS, 10 (1984), pp. 437–439.

[169] _____, *Algorithm* 624, ACM TOMS, 10 (1984), pp. 440–442.

[170] RIEMENSCHNEIDER, S. AND K. SCHERER, *Cardinal Hermite interpolation with box splines*, Constr. Approx., 3 (1987), pp. 223–238.

[171] SABIN, M. A., *The use of piecewise forms for the numerical representation of shape*, Ph.D. thesis, Hungarian Academy of Sciences, Budapest, 1977.

[172] SABLONNIÈRE, P., *De l'existence de spline à support borne sur une triangulation équilaterale du plan*, ANO–30, U.E.R. D'I.E.E.A.—Université de Lille I., 1981.

[173] _____, *Bases de Bernstein et approximants splines*, Ph.D. thesis, Université de Lille I., 1982.

[174] _____, *Interpolation by quadratic splines on triangles and squares*, Computers in Industry, 3 (1982), pp. 45–52.

[175] _____, *A catalog of B-splines of degree ≤ 10 on a three direction mesh*, unpublished manuscript.

[176] _____, *Composite finite elements of class C^k*, J. Comp. Appl. Math., 12 (1984), pp. 541–550.

[177] _____, *Bernstein-Bézier methods for the construction of bivariate spline approximants*, Comput. Aided Geom. Design, 2 (1985), pp. 29–36.

[178] _____, *Eléments finis triangulaires de degré 5 et de classe C^2*, in Computers and Computing, P. Chenin, C. Crescenzo, and F. Robert, eds., John Wiley, New York, 1986, pp. 111-115.

[179] SABLONNIÈRE, P., *Composite finite elements of class C^2*, in Topics in Multivariate Approximation, C. K. Chui, L. L. Schumaker, and F. I. Utreras, eds., Academic Press, New York, 1987, pp. 207–217.

[180] SCHOENBERG, I. J., *Cardinal Spline Interpolation*. CBMS Vol. 12, Society for Industrial and Applied Mathematics, Philadelphia, 1973.

[181] SCHOENBERG, I. J. AND A. SHARMA, *Cardinal interpolation and spline functions v. B-splines for cardinal Hermite interpolation*, J. Linear Algebra and Appl., 7 (1973), pp. 1–42.

[182] SCHOENBERG, I. J. AND A. WHITNEY, *On Pólya frequency functions* III, Trans. Amer. Math. Soc., 74 (1953), pp. 246–259.

[183] SCHULTZ, M. H., *L^2 multivariate approximation theory*, SIAM J. Numer. Anal., 6 (1969), pp. 181–209.

[184] SCHUMAKER, L. L., *Fitting surfaces to scattered data*, in Approximation Theory II, C. K. Chui, G. G. Lorentz, and L. L. Schumaker, eds., Academic Press, New York, 1976, pp. 203–268.

[185] ———, *On the dimension of spaces of piecewise polynomials in two variables*, in Multivariate Approximation Theory, W. Schempp and K. Zeller, eds., Birkhäuser, Basel, 1979, pp. 396–412.

[186] ———, *Spline Functions: Basic Theory*, John Wiley, New York, 1981.

[187] ———, *On shape preserving quadratic spline interpolation*, SIAM J. Numer. Anal., 20 (1983), pp. 854–864.

[188] ———, *Bounds on the dimension of spaces of multivariate piecewise polynomials*, Rocky Mountain J. Math., 14 (1984), pp. 251–264.

[189] ———, *On spaces of piecewise polynomials in two variables*, in Approximation Theory and Spline Functions, S. P. Singh, J. H. W. Burry, and B. Watson, eds., Reidel, Dordrecht, 1984, pp. 151–197.

[190] ———, *Triangulation methods*, in Topics in Multivariate Approximation, C. K. Chui, L. L. Schumaker, and F. I. Utreras, eds., Academic Press, New York, 1987, pp. 219–232.

[191] SCHUMAKER, L. L. AND W. VOLK, *Efficient algorithms for evaluating multivariate polynomials*, Comput. Aided Geom. Design, 3 (1986), pp. 149–154.

[192] SIVAKUMAR, N., *On bivariate cardinal interpolation by shifted splines on a three-direction mesh*, unpublished manuscript.

[193] STILLER, P. F., *Vector bundles on complex projective spaces and systems of partial differential equations* I., Trans. Amer. Math. Soc., 298 (1986), pp. 537–548.

[194] STÖCKLER, J., *Geometrische Ansätze bei der Behandlung von Splinefunktionen mehrerer Veränderlicher*, Diplomarbeit, Duisburg, 1984.

[195] STRANG, G., *The dimension of piecewise polynomials, and one-sided approx-*

imation, in Conference on the Numerical Solution of Differential Equations, G. A. Watson, ed., Lecture Notes 363, Springer–Verlag, Berlin, 1974, pp. 144–152.

[196] STRANG, G., *Piecewise polynomials and the finite element method,* Bull. Amer. Math. Soc., 79 (1973), pp. 1128–1137.

[197] STRANG, G., AND G. FIX, *A Fourier analysis of the finite element variational method,* in Constructive Aspects of Functional Analysis, G. Geymonant, ed., 1973, pp. 793-840.

[198] UTRERAS, F. I., *Positive thin plate splines,* Approx. Theory Appl., 1 (1985), pp. 77–108.

[199] ———, *Constrained surface construction,* in Topics in Multivariate Approximation, C. K. Chui, L. L. Schumaker, and F. I. Utreras, eds., Academic Press, New York, 1987, pp. 233–254.

[200] WALKER, R. J., *Algebraic Curves.* Princeton University Press, N.J., 1950.

[201] WANG, R. H., *The structural characterization and interpolation for multivariate splines,* Acta Math. Sinica, 18 (1975), pp. 91–106.

[202] ———, *The dimension and basis of spaces of multivariate splines,* J. Comp. Appl. Math., 12 (1985), pp. 163–177.

[203] WANG, R. H. AND T. X. HE, *The spline spaces with boundary conditions on nonuniform type-2 triangulation,* Kexue Tongbao(English edition), 30 (1985), pp. 858–861.

[204] WARD, J. D., *Polynomial reproducing formulas and the commutator of a locally supported spline,* in Topics in Multivariate Approximation, C. K. Chui, L. L. Schumaker, and F. I. Utreras, eds., Academic Press, New York, 1987, pp. 255-263.

[205] WHITELEY, W., *A matrix for splines,* unpublished manuscript.

[206] ŽENÍŠEK, A., *Interpolation polynomials on the triangle,* Numer. Math., 15 (1970), pp. 283–296.

[207] ———, *Polynomial approximation on tetrahedrons in the finite element method,* J. Approx. Theory, 7 (1973), pp. 334–351.

[208] ———, *A general theorem on triangular C^m elements,* RAIRO Numer. Anal., 22 (1974), pp. 119–127.

[209] ZWART, P. B., *Multivariate splines with non-degenerate partitions,* SIAM J. Numer. Anal., 10 (1973), pp. 665–673.